Videotext

The Coming Revolution
in Home/Office Information Retrieval

Efrem Sigel
with
Joseph Roizen
Colin McIntyre
Max Wilkinson

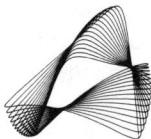

Knowledge Industry Publications, Inc.
White Plains, New York

Communications Library

Videotext: The Coming Revolution in Home/Office Information
Retrieval

Library of Congress Cataloging in Publication Data

Main entry under title:

Videotext.

  (Communications library)
  Includes index.
  1. Data transmission systems. 2. Television.
I. Sigel, Efrem. II. Series: Communications library.
TK5105.5.V52   778.59   79-18935
ISBN 0-914236-41-5

Printed in the United States of America

Contents

Illustrations

Preface

In the few short years it has been with us, videotext has already spawned its own terminology, a bewildering hodgepodge of brand names, acronyms, jargon and descriptive terms. Besides the brand names like CEEFAX, ORACLE, Telidon and ANTIOPE, participants in this field use generic terms like teletext, viewdata, videotex and videotext to describe the various technologies.

Our own preference is videotext. It combines two simple English words known to all and accurately describes the essence of the new service—the display of textual information, both words and numbers, on a video display screen.

This work would not have been possible without the timely efforts of my fellow contributors on both sides of the Atlantic, or without the assistance of many individuals and organizations who shared information with us. Special acknowledgement goes to the British Post Office for permission to use the photograph on the cover of this book, which is copyrighted by the Post Office.

Videotext: The Coming Revolution in Home/Office Information Retrieval is dedicated to the engineers, writers and editors around the world working to combine the technologies of computers and electronics with the art of creating the written word.

Efrem Sigel
November 1979

1

Introduction

by Efrem Sigel

It's probably not what Nicholas Johnson, former Federal Communications Commission member and a vehement critic of TV programming in the U. S., had in mind. In 1970 Johnson wrote a book entitled *How to Talk Back to Your Television Set* about how viewers could pressure broadcasters to put on better programs. But by 1979 thousands of TV owners were "talking" to their sets in a different way, pushing buttons on calculator-like keypads to instruct the television to display exactly what they wanted to see. What they were seeing, however, was not visual programming like "Upstairs, Downstairs" or the Olympics, but printed information—news summaries, financial reports, classified ads and descriptions of insurance policies.

As the decade of the 1980s approaches, new technologies and new economic forces are at work to turn that most familiar of appliances, the television screen, into a true information terminal. These technologies are coming from the electronics revolution in the two worlds of video and computers. The economic forces reflect the willingness of business (and some individuals) to pay high costs for information, provided the information is tailored to their needs and can be delivered nearly instantaneously.

There are as yet only glimpses in the United States of this new world of videotext, or TV-transmitted information, but developments are coming pell-mell in countries like Britain, France and Canada. There, information retrieval systems with names like CEEFAX, viewdata, Prestel, ANTIOPE and Videotex are moving from labs and research centers into homes and offices. By mid-1979 several experimental services had been announced in the U.S. as well.

How the TV set becomes an information terminal is simple enough to sketch. In one version videotext is simply a news ticker, displaying brief items of wide interest, such as latest news, financial reports, racing and sports results, weather information and airline or train schedules. These items can be "broadcast" on unused lines of the normal TV picture and displayed on any TV set equipped with a special adaptor to read the information.

Such a system is already in tens of thousands of British homes and public places, and could be in the U. S. by 1980, if various industry and government groups agree on technical standards. The key attributes of this approach are that the amount of information to be transmitted is fairly limited, and the user's role is strictly passive: he or she can watch or not watch, but has no say in the information transmitted.

In another version, dubbed Prestel in Britain, the TV set becomes a computer terminal in the home or office. If the set is equipped with the right electronic gear, viewers can order exactly what they want to see by punching a few buttons on a calculator-like keypad. The signals go over an ordinary telephone line to a computer, which responds by sending the information to just that home or office. What distinguishes this approach is that the amount of information available is potentially enormous (anything that can be stored on computer disks), and the method of transmission is what the computer people call "interactive." In other words, the computer tells Mr. Smith, "You can look at the following types of information: general, financial, travel, housing and employment." Smith says, "Financial," and the computer then offers more specific choices: "latest stock prices, brokerage house reports, economic projections, agricultural prices." Smith zeroes in on what he wants—e.g., the latest Merrill Lynch recommendation on IBM— and the computer displays it.

It is difficult to realize the seductive attraction of these videotext services until you see them print out information that interests you personally. Videotext displays are in up to seven colors plus white, allowing a full range of contrasts. Simple graphics, like maps or charts, are available, and some as yet experimental systems will be able to transmit still photographs.

Adaptors could be programmed to automatically receive and store certain information to be viewed when it is convenient:

thus, a much slimmed-down morning newspaper could actually be transmitted during the night, then scanned over morning coffee. One missing link at present is the existence of a reliable, low-cost printer that can be attached to the set. Once it is developed, it will enable the user to print out those pages he wants to save, to be stuffed into the attache case for the morning train ride or filed with the cooking recipes.

How this new technology will affect the television set is a provocative question to which, as yet, no answer is possible. Since the advent of commercial broadcasting in the 1940s, the television set has been the object of fascination and scorn by social critics. In nearly every industrialized country (with the exception of totalitarian ones, where the TV or radio receiver is an instrument of political indoctrination), broadcasting has become the pre-eminent medium of mass entertainment. The use of television for educational or information programming has made important strides, but it rarely constitutes more than a fraction of what is sent out over the airwaves. And this is true whether the economic basis of television is advertiser support, as in the U.S., or government subsidy, as in most European countries.

The overriding reasons for the emphasis on entertainment are both economic and technical. They hold true whether a country's economic system is capitalist or socialist. Television programming uses enormous chunks of bandwidth (portions of the radio frequency spectrum). This means that relatively few channels can be made available by governmental authorities for broadcasting. Whatever the nature of economic support for broadcasting, there will be pressure to use these scarce channels to reach the largest number of people. That inevitably means mass entertainment. Cable television greatly increases the number of available channels, but requires an enormous capital investment to lay the cable and hook up the individual homes.

In contrast, videotext, or informational television, is spartan in its use of the radio spectrum. Printed information for video display can be sent out in a tiny portion of the television picture, or over an ordinary phone line. It can also be sent out by microwave or an FM radio signal. In all cases, the sending facilities (telephone network, broadcast transmitter) are already in place. A fairly modest expenditure for an adaptor to the TV set is all that is needed.

The most ambitious type of videotext, involving computer storage of vast quantities of information (data bases or data banks, in computer jargon), of course requires big investments. But it is in this sphere that technology is moving most rapidly. The cost of computer storage is dropping by 35% a year, while the cost of performing a given computer operation in a microprocessor is falling nearly as fast. Storing millions of pages of text information in computers was once wildly impractical because of the cost. Today it is already practical for certain kinds of information, either information that is constantly examined and must be up to the minute (stock prices are an obvious example) or information that is infrequently examined but must be quickly available when needed (abstracts of scientific articles or an organization's personnel records are two examples). During the next decade, it will become economical to use computers for storing more and more of the published information used daily by individuals and organizations. As the process evolves, videotext services could become a logical way to transmit information to those who need it.

The purpose of this book is to show a variety of videotext systems in their early stages of development and application. It will quickly become apparent that the videotext phenomenon is worldwide. The British were the first to put videotext on the air through broadcasting, and the first to try a sophisticated information retrieval system using computers, the telephone network and specially adapted TV sets. The French are testing both types of services, and important research is also underway in Canada and Japan. The U.S. has so far lagged in research and development but individual U.S. companies like Texas Instruments and Zenith Radio Corporation seem ready to mass produce the receivers or components needed for videotext. Since the U. S. has by far the largest number of computers, computer terminals, telephones and TV sets of any country in the world, it seems safe to make the following prediction: if videotext is viable at all, it will spread wider and faster in the U.S. than elsewhere, even if it comes two or three years later.

2

The Technology of Teletext and Viewdata

by Joseph Roizen

INTRODUCTION

Teletext and viewdata are the current generic names for data transmission systems that utilize television channels, FM radio broadcasting, phone line circuits or a combination of the three, to bring information to viewers equipped with a properly modified color TV receiver.

Teletext is considered a one-way system, which piggybacks digital data on the normal television broadcast signal by inserting its messages in unused lines of the vertical interval. The home viewer with a normal television receiver sees only the regular TV program. A viewer with a teletext receiver has a control keypad that resembles a small hand calculator. By pushing certain buttons, the viewer can command the home receiver to shut off the regular TV program on the screen and to display instead "pages" of alphanumeric or graphic information that is being constantly updated by the TV studio. An index page (or pages) tells the viewer what information is available by page number. Selecting the appropriate page will bring the viewer such items as news headlines, sports results, weather forecasts, entertainment or shopping guides, and a variety of other public interest material. A basic teletext system limits the viewer to information that the TV studio's editorial staff decides to put out, and there is no real time method by which a viewer can request additional or different information.

Viewdata, on the other hand, is an interactive (two-way) system in which the viewer can request information from a computer data bank, typically using a phone line to the computer. Information is displayed on the screen of the home or office television set. In this way, the person can get up-to-the-minute stock quotations from one data base, airline or train schedules from another, local houses for sale from a third.

There can, of course, be variations on these basic themes. Teletext may employ only a few hidden lines in the television image for the transmission of a limited number of pages (often 100), with a reasonable access time for the viewer running about 10 to 20 seconds. However, a teletext system can also be expanded to full TV channel capacity, meaning that no regular program would be transmitted; the channel would be solely devoted to teletext. Under these conditions, thousands or even tens of thousands of pages can be sent over the system with reasonable access time.

For viewdata, it is possible to use other transmission channels than the telephone for the signals. The color television display terminal—equipped with a special decoder—may receive its information from an RF (radio frequency) transmission via a TV studio or an FM radio station. A buffer memory in the color TV receiver can accumulate the data being sent at whatever rate it arrives. When there is a full page of information in the memory, the viewer can transfer it to the display screen by punching the desired number on his or her keypad.

The return link from the user to the information source will usually be a telephone line. The viewer dials up a sequence of numbers indicated in the viewdata directory to get access to the selected data bank. Index pages then show the exact location of the information being sought.

It is possible to use a more elaborate return link. Either two-way cable TV methods or point-to-point microwave techniques will provide a wideband return channel; however, the cost of this is probably prohibitive and there is little need for such broadband capacity on the return side.

In the countries where they already exist (in some cases as experiments only), teletext and viewdata currently serve different purposes. Teletext is a public service intended to expand the utility of the home color TV receiver. The viewer need buy only a stand-alone decoder to attach to the set, or an integrally equipped teletext receiver, to have free access to the informational pages available from local TV stations.

Viewdata, on the other hand, imposes fees for access to its data bases. When users make a call to the computer, they enter an account number and are then billed special fees for the information, in addition to regular toll charges for use of the phone line. The mechanics of viewdata make it simple to levy a separate charge for each page of information, or for time spent communicating with the computer. The fees may vary according to the information requested.

The potential for a pure teletext system can be exploited most readily by television broadcasters. National networks, commercial networks, cable TV or pay TV companies can all offer expanded services to their viewers via such a system.

Viewdata applications are more likely to emerge in business, education, industry and government. In these areas, the ability of the system to transmit information rapidly on demand, and at reasonable cost, is the key to its acceptance.

There is little doubt that teletext and viewdata both have the potential to become significant, high dollar volume businesses in the next decade. However, each system will develop along different lines at a different rate. Viewdata's potential is more immediate because the prospective users already recognize its economic and operational advantages. In the United States, for example, it can enter the market on a purely competitive basis, with little concern for mandatory standardization or other regulatory action by local government agencies.

By contrast, an on-air teletext service in the U.S. must bear the scrutiny of the Federal Communications Commission, as well as various private standardizing groups like the Electronic Industries Association, Society of Motion Picture and Television Engineers and National Association of Broadcasters. It is eminently desirable to establish a single standard for teletext digital data signals to be transmitted over the air, so that a decoder of single design can interpret teletext signals anywhere in North America. This would parallel the current situation with regard to the NTSC color television system, which is the adopted standard for the U.S., Canada, Mexico and Central America.

Standardization takes time, perhaps one to two years of field testing, evaluation, committee action and FCC ratification. Until that happens teletext can only be a limited experimental service that holds the promise of future developments.

Moreover, developing teletext in North America has an economic as well as technical dimension. Broadcasters must find a way to be compensated for the cost of putting teletext pages on the air, since they will be losing advertising revenue if there is a decline in the number of viewers watching regular programs.

Outside of North America, teletext and viewdata services can be started at the pleasure of centralized telecommunications or broadcasting authorities. This simplifies the problem of persuasion or marketing, but there are still vested interests that are bound to oppose these new services either because they cost too much or because they threaten to compete with existing offerings.

HOW TELETEXT WORKS

The basic teletext technique is as follows:

1) The information, consisting of alphanumeric or graphic images, is encoded in a bit stream of digital data at a transmission rate that the television system can properly handle. This rate will vary according to which color TV system is in use in the country.*

2) The encoded digital data is inserted or multiplexed onto the TV signal waveform in such a way that it is located on unused lines in the vertical blanking interval. This is the period during which the scanning of the television picture begins again. Physically, the lines are located at the top of the TV screen, on the part of the tube hidden by the cabinet. They will not be visible on a properly tuned TV receiver. Hence the presence of the digital data or teletext signals is not noticed by a viewer watching a normal program.

---

*NTSC, PAL and SECAM are the three noncompatible color TV broadcasting systems. NTSC stands for National Television Systems Committee, PAL for Phase Alternate by Line and SECAM for Système Electronique Couleur Avec Mémoire.

3) The teletext signal can be detected by a special decoder that is either a separate accessory to the color TV receiver, or actually built into it. In either case, the teletext decoder circuitry can accept the digital data, store one or more pages in a buffer memory and display these pages on the screen as directed.

It should be stressed here that the teletext decoder, incorporating a microprocessor or "chip," is the key to predictions of a vast market for the technology. By producing this component in large quantities, electronics manufacturers like Texas Instruments can presumably make the cost very low, as they have done with the pocket calculator. The teletext decoder can also be one feature of a low-cost home computer that attaches to the TV set, permitting the users to call up information from outside as well as to perform their own calculations, create their own programs and store their own information.

4) When the viewer punches the number of the desired page on his control keypad, the buffer memory containing that page is kept in a "hold" condition. The page is then transferred to the color TV screen via a character and graphic generator, which is part of the teletext decoder circuitry. The page remains on the screen until a replacement page is transmitted, or until the viewer selects a new page.

There are several modes of teletext operation, including:

- TV program only with no teletext visible.
- Teletext only with no program material visible.
- TV program with a teletext overlay on all or part of the picture.
- TV program with a window in the picture through which teletext appears constantly. (This is called "boxing" in the British system.)
- TV program with provision for teletext override in an assigned window area should there be some significant event that the viewer wants to know about.

While broadcast teletext has thus far made use of a few unused lines of the vertical interval, it is not limited to that application only. Teletext signals can be applied to any number of the active television lines in the picture area or even all of them, thus filling the full television channel with data rather than program material.

Full channel teletext could be used by common carriers, by cable TV companies on an otherwise unallocated channel, or by pay TV operators and even broadcasters when the regular programming is off the air. As an example, a TV studio that goes off the air at 1:30 a.m. and doesn't start up again until 5:30 a.m. could transmit full channel teletext signals for four hours each night. These transmissions could contain many thousands of pages of data that are directed to special teletext receivers.

The coding of the transmission could limit access to certain blocks of pages to receivers with specially designed decoders. In this way, groups like doctors, attorneys, law enforcement officials, travel agents or any others, could receive pertinent information dedicated to their special needs. The memory bank in the set would store the information until the viewer looked at it, or the information could be processed by a printer attached to the TV set. Each morning the set owner would have a printout of the latest news in his or her field that had been transmitted during the night.

THE TELETEXT SYSTEM

Teletext transmitted over an active broadcast channel is intended as a supplementary service to viewers. It can bring them instant access to items ranging from news bulletins to the day's horoscope.

To provide this service the TV studio must create an editorial department that prepares teletext pages dedicated to specific topics. Editors, using usual news sources (UPI, AP, Reuters, etc.), scan the available information and enter it into the teletext system via a special keyboard that transforms the information into the digital code which will be transmitted. The editor composes a page on the keyboard and sees the result on a color TV monitor, where errors can be corrected or new data inserted. The digital data representing that page is put in a computer memory, where it is stored until it is no longer useful.

While current, that page in the computer is transmitted on a repetitive basis with all of the other available pages. The full sequence of these pages comprises the "magazine." How long a viewer may have to wait to see a desired page is based on the re-cycle time of each page; i.e., if there are 100 pages of information and a new one is transmitted every quarter of a second, then the maximum wait for a desired page is 25 seconds. The average wait, however, is only 12½ seconds.

Access time, a very important parameter in teletext systems, is actually dependent on four factors.

1) The number of television lines dedicated to carrying the information to the viewer. The more TV lines allocated, the faster the access to a particular piece of information.

2) The number of pages of information the teletext service wants to provide. The more pages, the slower the access time.

3) The digital bit rate used to transmit the information. A low bit rate will take longer to send the digital data over the available TV lines, but it will give a very reliable error-free result; a higher bit rate will speed up access time at the expense of the error rate. The maximum bit rate is also limited by the bandwidth of the TV transmission system. European teletext systems can use higher bit rates than those in North America because TV channels in Europe are wider.

4) The amount of memory in the decoder. If the memory holds only a single page, the access time to any page is governed by the three preceding factors. However, with a multiple page memory, only the access time to the first page may be relatively long. Every succeeding page will have a seemingly shorter or almost instantaneous access time, since these pages will get stored in the memory while the viewer is scanning the first one.

As an example, suppose before leaving for the office the viewer wants five pages of information as a morning update on the latest news, local weather, football results, stock quotations and traffic conditions. He or she consults the index for the appropriate page numbers, then enters each one in succession. The teletext decoder will take average access time to show the news page, but other pages (with weather, football, stock and traffic details) will be going into the memory while the viewer reads the news. By simply pushing a button for the next stored page after the news has been absorbed, the viewer will receive the already stored weather page almost instantly. Each subsequent page would also be available quickly.

In general, it can be assumed that digital memory will continue to get less expensive at a rapid rate. The teletext decoders of the future will have more than single-page memories. In addition, other schemes to cut access time are being developed or proposed. One approach is to have priority pages, which are repeated more frequently than those of lesser interest. In that case the decoder has the opportunity of grabbing that page several times in the paging sequence, thus providing faster access.

THE TELETEXT DISPLAY

While teletext may appear on either a black and white or color TV receiver, it is assumed that for all practical purposes a color set will be used. The color combinations available are composed of the primary (red, green, blue) and secondary (yellow, cyan, magenta) colors seen on a normal color TV receiver. In addition, white and black are available. Color increases the scope of information that can be conveyed because the various hues can be intrinsic carriers of additional data. For example, a stock market report page could show stock quotations in red, green or white, the color indicating if the price is down, up or unchanged.

Teletext systems also provide for varying character size, flashing of certain information to attract special attention and the conceal-ment of page sections until they are required to appear.

The selection of the number of characters per row and the number of rows on a full screen is also variable, but there are limits imposed by the visibility required. In Europe, there are more lines in the picture (625) and more spectrum bandwidth on the TV channel. The teletext systems operating in Britain and France can accommodate up to 40 characters per row and a maximum of 24 rows per raster (screen). This produces 150 to 200 words of legible text at normal viewing distances. In North America, with fewer picture lines (525) and a narrower transmission channel, one display that has been tested consists of 32 characters per row on a maximum of 21 rows per raster. Other displays provide up to 40 characters per row, with about 20 rows. It should be pointed out that outside of the United Kingdom, these display standards have not yet been officially adopted and are the subject of consideration.

As stated earlier, the bit rate for the transmission of the digital data is also variable within limits. The U.K. teletext system uses 6.9 megabits per second, the French system 6.2 mbs. and the proposed North American may end up with something between 3.5 and 4.0 mbs.

Last but certainly not least is the question of which unused vertical interval lines are to be dedicated to broadcast teletext. Television images in Europe consist of two fields, each containing 312½ lines; the two fields together make up the 625-line picture. The two broadcast services in the United Kingdom, the BBC and the Independent Broadcasting Authority (IBA), have jointly agreed upon a standard that allocates lines 17 and 18 in field one and lines 330 and 331 in field two for the public teletext service. As a result, a viewer in the United Kingdom can receive teletext transmissions on the same decoder when tuned to either BBC or IBA stations.

In France, which uses the SECAM color system, lines 16 through 20 and 329 through 333 are applicable to teletext services. In North America, the currently unused lines in the vertical blanking interval are 10 through 16 and 273 through 279. Some of these are not easily usable for technical reasons, so it is likely that lines 15 and 16 and 278 and 279 will be selected for teletext signals. The final decision will be made by the FCC after field trials, underway in 1979.

These tests will assess the impact on the teletext signal of propagation (how topography and geography affect transmission), multi path (a phenomenon caused by the TV signal reaching the home receiver both directly and indirectly, e.g., bouncing off mountains or tall buildings), fringe area reception (how signals will be received outside a specified radius from the transmitter) and error rate and efficiency (quality of the image as affected by distance and transmission speed).

All these technical problems, while present in any TV broadcasting system, are more critical for videotext than for normal programs. A viewer may not even notice a slight error in the regular TV picture, or can ignore the bit of snow on the screen. But if a technical problem causes the number 3 in a teletext display to come out on the screen as a 2, the whole value of the information may be destroyed.

Hence, both tests and the promulgation of technical standards must precede the manufacture of teletext decoders that can be built into (or attach to) any color TV receiver.

HOW VIEWDATA WORKS

The technology of viewdata systems is more complex and costly than that involved in broadcast teletext. But because such systems do not involve public broadcasting to a universe of millions of homes, they do not run up against the same need for standardization of hardware and programming. A number of noncompatible viewdata systems could exist in the same country, each making use of the telephone network or other common carrier facilities for transmission.

The essential elements of a viewdata service are:

1) A large computer that can store many thousands (perhaps even millions) of pages of textual information.
2) Computer programming (software) that permits the accessing and rapid retrieval of specific items of that information, and the billing of customers who use the system.
3) Transmission lines for sending information back and forth between the customer and the computer; these lines can consist of the public telephone network, a cable television system with two-way capabilities or special microwave facilities.
4) Display and retrieval terminals. These can be color TV receivers with a decoder attached to translate digital signals into the TV display, or modified computer terminals capable of color display. As with the teletext decoder, the microprocessor that can be manufactured in high volume is essential to a reasonable price. For use with phone lines the terminal must

contain a modem that converts an analog telephone signal into digital form for display. The retrieval device may be a simple calculator-like keypad with buttons for numbers 1 through 10, or a full typewriter-like unit.

Both the scope and cost of a viewdata service are far greater than for broadcast teletext. Setting aside a large computer for the service means a substantial capital investment. Creating the computer programming for two-way communication between customer and computer also requires a large initial cost, plus continued maintenance and improvement of the software. Finally, publishers or other organizations must spend money to create and edit the information that goes into the system; information that requires continual updating will of course be most expensive to maintain.

While the emphasis in teletext services has been on information with the broadest appeal, such as news headlines and sports results, viewdata systems can include much more specialized information, such as detailed financial data on companies or industries, information on the products or services of companies, classified ads and directory listings. Organizations can use the technology for communicating to their own employees or agents, using special computer codes that restrict access to their portion of the data base.

Although the regulatory issues are less evident with viewdata than with teletext, they can hardly be ignored. In the United States, it would appear that viewdata does not come under the purview of the FCC any more than a computer timesharing service would. However, the cooperation of telephone companies or other common carriers is essential to the creation of any nationwide network. In European countries, where a government agency runs the telephone/telecommunications network, the decision to install a viewdata service becomes, in effect, a political one. Even so, such a service does not compete with the telephone company's main business; it actually increases phone usage. In contrast, a teletext service may mean reduced viewership for TV programs created by the broadcasters.

VIEWDATA DISPLAYS AND TECHNICAL PROBLEMS

The display of viewdata information on a screen may or may not be identical to the broadcast teletext display, depending on the country. In the United Kingdom, both the British Post Office and the broadcasters have settled on a format of 40 characters by 24 rows; the transmission rate is identical and so are the colors used. In the United States, broadcasters must use a standard approved by the FCC, but manufacturers are free to make any computer display equipment they wish for non-broadcasting systems.

Some of the technical issues to be resolved in viewdata systems include:

- Will the information be displayed on existing computer terminals, modified TV sets or both?
- How large a storage capacity will the computer used in the system have, and how many users can it accommodate simultaneously? If the computer is not powerful enough, too many calls at the same will cause circuits to be overloaded; customers will experience frustrating delays in getting responses to their requests for information. If the computer is too powerful for the amount of activity, response time will be satisfactory but the costs too high.
- What communications system will give the best service? Use of the public telephone network would make a viewdata system universally available, but home users might be reluctant to tie up the only phone line in the house. Countries where the telephone company has a backlog of installations to make will find it difficult to offer very comprehensive viewdata services. Use of cable television systems in North America would mean plenty of spare channels available, but such systems reach only a fraction of households and businesses. Constructing special microwave facilities would be expensive; satellite transmission to rooftop antennas is technologically feasible, but some years away as a practical matter.

SUMMARY

By mid-1979 the technology for videotext was well understood, and systems were being installed in a number of countries. The technology involves the display of textual information and simple graphics on a TV screen, using a special decoder. In a broadcast version, often called teletext, the signal can be sent in unused lines of the regular TV picture, but capacity is limited to about 100 pages of text. In the non-broadcast version, often called viewdata, the capacity is limited only by the size of the computer used. Transmission by means of the telephone network or other two-way channel permits the consumer to select just what information he or she wants to see. Both systems have technical problems still to be resolved, but the more important problems are economic and psychological: Can the systems be made to pay for those who must invest in their creation? Can individual and business users be persuaded to use information in this entirely novel way.

There is no way to answer these questions in the abstract. Videotext is already moving into the arena of the marketplace, where its fate will be decided by suppliers and customers. The following chapters will describe the state of systems in various countries around the world.

3

Teletext in Britain: The CEEFAX Story

by Colin McIntyre

Britain is a good two years ahead of the rest of the world in introducing videotext services like CEEFAX and viewdata. The two television organizations — the British Broadcasting Corporation (BBC) and Independent Television (ITV) — have been transmitting broadcast teletext for more than four years, while 1979 saw the full-scale launch of the British Post Office's on-line system of viewdata, called Prestel. Both versions of these data information services are British developments.

The quick progress from the drawing board to a public service seems to reflect both the pragmatic way in which the British approach such inventions and the informality of decision-making by government authorities. Once the BBC's director of engineering first saw an experimental version of CEEFAX (the BBC's trade name for its own teletext system), it was a short step to deciding that the development constituted broadcasting and thus fell within the BBC's charter. The Minister of State for the British Home Office, which supervises broadcasting, had no objection. Subsequently, the medium was closely examined both by Lord Annan's Committee on the Future of Broadcasting and in a government White Paper. Both endorsed that first decision. After a two-year pilot trial, broadcast teletext became a full legitimate service in November 1976.

The launching of teletext has obviously been helped by the status of the BBC as an independent public corporation and by the overall supervisory role of the IBA (Independent Broadcasting Authority) on the commercial television side. In the case of the telephone viewdata system, rapid advance was made possible by the present constitution of the British Post Office, with its virtual monopoly of all communications in the country and its financial independence from government.

These conditions allowed the broadcasters and the Post Office, in amicable step with the government of the day, to press ahead very quickly from the first developmental stages: no lengthy constitutional arguments, no time-consuming commissions of enquiry, no references back and, above all, no buck-passing from one authority to another. Teletext in particular benefitted from the fact that the Minister of State at the Home Office responsible for broadcasting was Lord Harris of Greenwich, a former journalist, and that his successor was Lord Boston of Faversham, another journalist and an ex-BBC producer.

Confident decision-taking has also meant that a minimum number of organizations have become involved. Newspapers, which might have been expected to demand a say in the matter, are not allowed under British law to become directly involved in broadcasting, except as minority shareholders in commercial television and radio. Perhaps also, beset by many problems of their own in meeting the demands of the new technologies, they showed little interest in broadcast teletext. Newspapers are playing a more active part in the much more extensive viewdata telephone system (Prestel), which in theory could eventually connect a home television set with every computer in the world, and which in practice begins with some 100,000 "pages" or frames of information.

Another reason for quick progress in the development of teletext and viewdata was the early involvement of the television manufacturing industry—the firms that make actual television sets. In the case of teletext, not only did the two rival broadcasting organizations get together to pool their research findings and produce one United Kingdom Teletext Standard, but technical members from BREMA (the British Radio Equipment Manufacturers' Association) were brought in to help define the final transmission standards. This was fairly unique, as hitherto the broadcasters had tended to announce the parameters of their transmissions, and then leave it to the set makers to meet these standards and produce the necessary receivers. The set makers were farsighted enough to design for the kind of signals appropriate to the rapidly developing electronic world of the 1980s.

What emerged therefore was a truly joint Technical Specification

CEEFAX 102   Tue 5 Oct 11:11/30

BBC news

Mrs Margaret Thatcher has warned the
Conservatives, whose Party Conference
starts today, to get their election
machinery ready.

She said an election could come at
short notice.

Delegates at Brighton will be debating
four major topics today; the economy,
education, Northern Ireland, and
immigration.

CONSERVATIVE PARTY CONFERENCE - PP190/1

A typical CEEFAX "page" showing the latest news. The page (102), date and time appear at the top of the screen. ©BBC, reproduced with permission.

for Broadcast Teletext (September 1976), published by the BBC/ IBA/BREMA. For its on-line viewdata system, the Post Office agreed to accept the broadcasters' patterns for screen display: the 960 character-spaces defined for broadcast teletext (40 characters across by 24 rows deep), and the same colors based on the red/green/blue of broadcasting to make the seven colors of on-screen display: red, green, blue, magenta, yellow, cyan and white. (Black makes an eighth color.) By 1978, many of the same television manufacturers who were producing teletext sets had also met the Post Office safety and technical requirements for Prestel receivers, and other sets were under evaluation. (Prestel will be discussed fully later in this chapter and in Chapter 4.)

DEVELOPMENT OF CEEFAX

The Beginnings

The history of the development of broadcast teletext began with the efforts of the then BBC director of engineering, Sir James Redmond. He was looking for ways of producing subtitles for the deaf and hard of hearing, the so-called "closed captions" system, which would involve subtitles to be seen only on the television screens of deaf people who bought an extra piece of equipment — a decoder. These subtitles would not intrude, could not even be seen, by those who did not have a decoder or did not dial the service.

Considerations of subtitles obviously apply to language minorities, as well as the deaf. In the case of Britain there is reason to subtitle in Welsh, Gaelic, Hindi and Gujerati, among other languages. In the U.S. or Canada, programs could all carry Spanish or French closed captions; in South Africa, both English and Afrikaans. Again, in countries such as Belgium or Denmark, where several television services from across their borders can be received, the transmitting country might find an incentive to offer closed captions.

The then head of engineering research at the BBC, Peter Rainger, who had been writing and thinking about the possibilities since 1970, put together a number of his ideas and set up a research team under a veteran engineer, Stan Edwardson. Computers were coming down rapidly in cost, and a number of bits of allied technology — such as the Read Only Memory — began to seem particularly appropriate to Rainger's ideas. Advances had also been made in data transmission. Thus, in a comparatively short space of time, during 1971 and 1972, the BBC team at Kingswood Warren, some 20 miles from London, had come up with the elements of CEEFAX. Its name came from "see facts," or perhaps BBC-facts, and once understood was fairly unforgettable.

Meanwhile, another team, working independently at the Independent Broadcasting Authority, was producing its own version, to be known as ORACLE (Optional Reception of Announcements by Coded Line Electronics). A paper by Peter Hutt of IBA was published in late 1972 outlining ideas very similar to those being developed at Kingswood Warren. Both broadcasting organizations took steps to coordinate these advances, in the process accepting technical ideas from British TV set manufacturers.

By mid-1972, Redmond was convinced that the final product from his own laboratories was practical and fell within the BBC's mandate of broadcasting. He and Rainger had the imagination to realize that what had been thought of as a solution to the subtitling problem was indeed something quite else, something much bigger — in fact, a completely new medium for broadcasting.

In addition to subtitles, CEEFAX could be used for every possible kind of news and information: from sports to politics, from foreign crises to home prices, from travel to weather maps, and from rates of exchange to joke-of-the-day. And also for telling the time. (Every page of teletext carries the actual time in the top right-hand corner. It appears as a six-figure data clock, in yellow, split as 16:22/15 [hours, minutes, seconds]. The time is based on an extremely accurate clock at the National Physical Laboratory in Teddington — a device often described as an atomic clock, though it is actually a rubidium vapor oscillator.)

Test Broadcasts

The first test pages of CEEFAX were on the air by mid-1973. Although these were mainly engineering test pages or messages of "The quick brown fox jumps over the lazy dog" variety, this progress from the laboratory to full on-air transmissions over a network was surprisingly quick. Once it was clear that teletext worked technically, and that signals could be received almost anywhere that a good television picture reached, the director of engineering was eager to translate his development into a real public service. He found the necessary money in a contingency fund, and persuaded his colleagues on the Board of Management to start on a limited on-air experiment with real news and information.

As a longtime journalist in the BBC, I was invited to be the first program man associated with this project, which was still housed in the BBC's Engineering Division. Offering a journalist a completely new way of publishing what he has written is rather like offering a grafitti expert a huge new blank wall, and then giving him seven different-colored pots of paint to use on it. Beginning at first on a research visual display unit (VDU), which simply cut punched-tape to be stored in a 30-page core store memory, was a trifle laborious, but nonetheless exhilarating. Within days we had an outline "magazine" to demonstrate to the BBC's Board of Governors and directors, Home Office broadcasting officials and telecommunications experts.

The BBC made up its mind to go ahead, and I was allowed to recruit a team of eight—four assistant editors (called sub-editors) and four young research assistants (in effect, trainee journalists-cum-reporters)—and told to get a pilot service going. The government gave the BBC and IBA a first qualified approval for a two-year pilot trial, and CEEFAX began its first broadcasts of live news and information a few days later — September 23, 1974. The date chosen was the start of an International Broadcasting Convention meeting in London; the convention was opened by the Duke of Kent, who was thus one of the first viewers to see the service operating.

Looking back, it all seems rather slow and amateurish. Tapes had to be punched on the single VDU and then carried through an engineering area to be fed into our memory store. We had 30 pages to fill, although some of them were still needed for engineering test signals, or for demonstrations of the CEEFAX graphics capability. The next steps were to buy and program a computer and to build a special newsroom on the seventh floor of the BBC's Television Centre.

Start of Full-scale Operations

By early 1975 CEEFAX was really in business. The computer was nicknamed "Esmeralda." An off-the-shelf Computer Automation

ALPHA LSI/2 from the United States, it cost somewhere around $25,000. In fact, the BBC's original transmission system for CEEFAX — comprising a computer, six input units, equipment to combine the data with the ordinary television picture, indeed everything needed to provide a two-network nationwide service— cost less than £200,000 ($400,000). That is roughly the cost of two color cameras in a TV studio, or an expensive television opera production.

Landmarks in CEEFAX history include our start of parallel transmissions on the BBC-2 second network, the visits by members of Lord Annan's Committee, the establishment of the first outpost terminal at Broadcasting House, five miles away, and the full go-ahead from the government on November 9, 1976. Along the way were some unofficial landmarks, such as the visit to the CEEFAX newsroom of a toddler who covered every keyboard in white eradicating fluid, or the time a coffee spoon and bent cigarette pack held one of the computer's boards in place.

Almost all the CEEFAX team of journalists came from outside the BBC, in order to guarantee a fresh approach to this new arm of broadcasting. They came from the wire services or from local newspapers or, in the case of the young researchers, virtually straight from university. They were a young team, with the exception of the editors, all under 40 and most in their twenties. It was a sign of established success when, within a short time, original CEEFAXers left to join a national newspaper or an international news agency, while the researchers got promoted to other jobs.

THE CONTENT

Perfecting and establishing broadcast teletext technically was only half the battle, for in the last resort it is the use to which a new technology is put that really matters. What is it that teletext provides that other media do not, or do not provide as well?

Convenience

One answer that emerged in the early days was to think of teletext as *printed radio* — to see teletext as the words of radio news and other informational programs (travel reports, sports commentary, even cooking recipes) simply reproduced in a handy, constantly accessible form for viewing on a home television screen. That is certainly part of the story: the *convenience* of having news and information when you want it, and in the order of your choice. You get your news bulletin when you happen to come into the house, or at a convenient break in your evening's viewing, when you return from dining out or going to the cinema, or before you go to bed. And you only read those items that interest you, as an individual. If you are a golf-playing commuter who invests in the stock markets, you will remember the page numbers for golf, weather, travel and stock prices. If you are a society lady who loathes sport, you need never see another baseball result. If you are a housebound senior citizen, you can keep up with everything from reviews of opera and ballet to the films coming up soon on TV, from horoscopes to sports quizzes, from shopping hints to road reports on new cars.

For the first time it is the individual viewer who makes up his own custom-built news bulletin, who reads the items in his own order of importance, not one imposed upon him by a newscaster or newsroom. This convenience factor is high, and equally impossible to measure.

Speed

The other great virtue of teletext is *speed*, and the absence of deadlines or fixed bulletin-points. News is transmitted as it becomes available, not only world-shattering items, but ordinary items that might happen to be important to a particular viewer. A newsflash, which the teletext system can actually intrude into the ordinary picture, can be broadcast within minutes of it reaching the newsroom — as fast as an assistant editor can type. One page of the CEEFAX service is set aside for watching the regular network program, with provision for inserting late-breaking news as it happens. The photograph on the next page shows what this looks like.

During the 1975 Wimbledon Men's Tennis Final, the CEEFAX sub-editors prepared two separate news flashes and stored them in the computer. One read:

Men's Singles Tennis: 5-5 in the fourth
set, with Ashe leading by two sets to one.
The other said:

Ashe wins Wimbledon: 6-1, 6-1, 5-7, 6-4
The CEEFAX staff was watching the game on television; when the

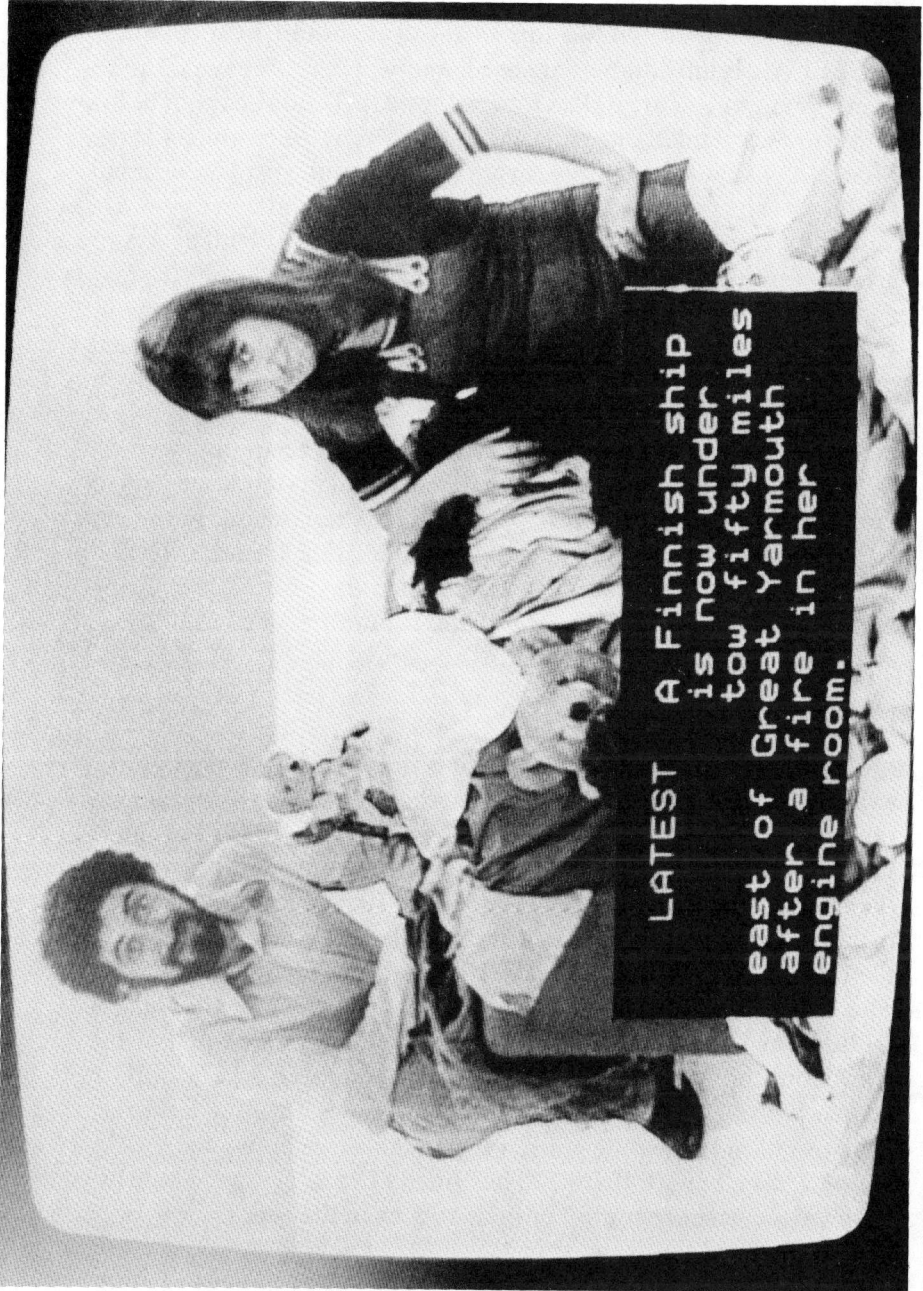

The ability of CEEFAX to cut through the ordinary TV program (in this case, "Play School") with a news flash.©BBC, reproduced with permission.

ball hit the net on Jimmy Connors' side, and before the umpire had had time to pronounce "Game, set and match to Ashe," the computer had delivered the correct result on both BBC networks.

CEEFAX is unbeatable for speed, and will become even more so as other computers are linked into the system. The day cannot be far off before Esmeralda or her successor is married to computers at the Stock Exchange or London Airport. The airport display terminal could be connected in such a way that aircraft arrivals are automatically displayed on a given page of CEEFAX; all delays of arrival and departure are automatically fed into the teletext computer. Instead of trailing out to Heathrow Airport to meet your rich Australian aunt, you will be able to see that her plane is delayed overnight in Bahrein with engine trouble and stay comfortably in bed.

In editorial matters, there will always have to be an intervention by the broadcaster, as editor and publisher. But in simple matters of fact, for which an information provider is the responsible authority (stock prices, lottery results, train timetables), the information can go straight from computer to computer in data form; after being appropriately repackaged by the teletext computer, it can emerge as a CEEFAX page — untouched by human hand.

Choosing the Content

Deciding what features to include is not as difficult as might be expected, even though it may sound arrogant to say so in an age that relies so heavily on market research, polling and consumer interviews. The BBC has, after all, been broadcasting radio for more than 50 years and high definition television since 1936. And it maintains a constant 50% share or more of the television audience in the face of the efforts by its richer rivals in commercial television. The BBC hears from its audience, to the tune of more than half a million unsolicited letters a year. The size and composition of the television audience is well established: there are some 3 million people who will watch horse racing on television on a Saturday afternoon, and some 10 million who regularly choose the BBC's main evening news bulletin, "The Nine O'Clock News." The BBC also knows there are some 3 million anglers in the country, but only several hundred thousand skiers.

From the start, CEEFAX was seen as entirely broadcast-related, as an integral part of the BBC's broadcast output on both radio and television. We did not seek to find new and radically different items to put on CEEFAX. Already a vast range of topics are dealt with by broadcasters, beyond the obvious bread-and-butter items of news and weather and sports: anything from religion to railway lore, from chess to computers, from photography to puzzles. In fact there is almost no subject that is not in some way dealt with on the air, and in which CEEFAX — as broadcasting — cannot find a legitimate interest.

Indeed, the problem is to try not to do too much, but to identify those features that CEEFAX can do best, or better than existing media. There is little point, for instance, in printing train timetables, which would occupy vast acres of CEEFAX space and can be done just as well elsewhere. But it *is* essential to be able to report major delays, breakdowns, cancellations or lightning strikes, which can affect the traveller. (Avoiding one traffic jam or one missed train a week can pay for the weekly rental of a CEEFAX-equipped set.) In terms of the BBC's program information, it would be wasteful to carry all the details you would expect to find in *Radio Times* or *TV Guide.* But it is a real service to the public to be able to announce last-minute changes, explain why schedules are running late or give advance notice about stars who have accepted a late booking to appear on a talk show. It is the *immediacy* of teletext that is hardest to explain in writing or lecturing about it. A full understanding only comes with the ownership of a set, over a period of time.

Interaction with the Viewer

If CEEFAX makes a spelling mistake or mis-labels a public figure, a viewer inevitably telephones within four or five minutes. The viewer knows that behind that mistake on his television screen is a live journalist. He will call in the knowledge that the journalist is still there and can do something about the error. With all respect to broadcasting colleagues, the public does know that as the last reels of "Midnight Movie" crank out, probably the only people still on duty in an otherwise closed and darkened television studio are a lone engineer and an announcer standing by to say goodnight. But in the case of CEEFAX, the viewer is aware that someone is there, still updating as the bewitching midnight hour strikes. If the viewer's complaint is valid, the CEEFAX staff member may say, "You are

A teletext viewer operates the keypad used to select the page she wants to watch. An adaptor made by Labgear hooks up to any standard color TV set. Courtesy BBC.

quite right, madam. Just keep watching and we shall change it."
And the change takes place in front of her very eyes. If this is not
two-way interaction, what is?

Flexibility

The other great strength of teletext as a broadcasting medium is
that there is no capital tied up in the program content itself. It is not
a matter of junking large cans of film or video tape if later events, or
public taste, decree that these prepared programs are unwanted; it is
not a case of sacrificing already-recorded programs that prove
unpopular. On CEEFAX, if a page or section deserves death, we
simply type over it. CEEFAX has great flexibility in meeting the
public's information tastes. If it turns out that what the public wants
is not the present mix of news, sports and finance but 800 different
knitting patterns, or 800 different recipes for soup, we could
probably provide this by the next day, after a mammoth night of
typing. We would have some mightily bored journalists, but
presumably a gratified public.

The CEEFAX BBC-1 and BBC-2 indexes as of early 1979 are
shown on the following pages. BBC-1 is the British Broadcasting
Corporation's main network and is on the air for some 18 to 20
hours. Thus, it was obvious that BBC-1 would have to carry the
principal CEEFAX service — the news, sports, business, travel and
weather. BBC-2, the second network, is on the air fewer hours per
day and does not really get going until late afternoon (except for an
occasional children's program or race-meeting). An early editorial
decision was to use BBC-2 CEEFAX as something more of a weekly
or biweekly magazine, as opposed to the daily newspaper orientation of BBC-1 CEEFAX. It was used for background news (the state
of the parties during an election campaign, maps and supporting
diagrams during a military conflict, or biographies of people in the
news). A correspondence with BBC-1 was maintained, so that the
equivalent numberings on BBC-2 CEEFAX listed the sporting or
business calendars, or long-term travel news and long-range
weather forecasts.

But none of this is necessarily permanent. BBC-2 might easily be
used for regional or local information, for education or other
specialized purposes. The complete flexibility of teletext must
always be emphasized, with everything able to be changed in the

Ceefax on BBC1

Contents: Page 100 (BBC-1)

NEWS		FINANCE		SPORT	
Headlines	101	Index	120	Headlines	140
News in detail	102-116	News and Reports	121-126	Sports News	141-159
News Diary	117	Market Reports	127-129	CEEFAX provides a rapid	
People in the News	118	FT Index	130	service of news, results	
Charivari - a lighter		Stocks and Shares	131-133	and background.	
look at the News	119	Exchange Rates	134-136	CEEFAX Sports Specials,	
		Commodities	137,138	covering major sporting	
				events, begin on 151.	

FOOD GUIDE		ENTERTAINMENT		WEATHER AND TRAVEL	
Headlines/Index	161	Today's TV - BBC-1	171	Headlines/Index	180
Shopping Basket	162	BBC-2	172	Weather Maps	181
Meat Prices	163	ITV	173	Temperatures	182
Fish Prices	164	Radio highlights	174	Temperatures	183
Vegetable Prices	165	Films on TV	175	Travel News	184-189
Fruit Prices	166	Top Twenty	176		
Recipe	167	Theatre	177		
Farm News	168-169	Opera/Ballet	178		
		Viewers' Questions	179		

NEWSFLASH 150	ALARM CLOCK PAGE 160	SUB-TITLES 170
Turn to this page to watch television programmes - when something important happens a NEWSFLASH will appear on the picture.	This page can change every minute. It can also be used as a silent alarm clock. Turn to page 160 for instructions.	The BBC is experimenting with various ways of sub-titling programmes. This page shows how sub-titles could look.

LATEST PAGES 190	OTHER PAGES	WANT TO KNOW MORE ?
As each new page is put in the magazine it is also put on 190 where it alternates with a news summary.	News about CEEFAX 191 Engineering tests 197-198 FULL INDEX A - F 193 G - O 194 P - Z 195	Write to: CEEFAX Newsroom (7059) BBC Television Centre, LONDON, W12 7RJ

Ceefax on BBC2

One of the great advantages of CEEFAX is its flexibility.
Old pages can be removed and new ones put in virtually instantly. For
this reason, CEEFAX pages can vary from day to day as new ideas are
tried out. To help you find your way around, there is a full index
towards the end of each magazine — pages 193-195 for Magazine 1 (BBC-1)
and 293-295 for Magazine 2 (BBC-2).

time it takes to type it. New pages, new themes, are being added or tried out all the time. For special events (elections, great sporting occasions, major disasters) whole sections of the magazine are cleared and devoted to the particular occurrence. A mention on the main index page, and then the use of a further index, can produce a magazine-within-a-magazine, again, almost overnight. At Christmas and on St. Valentine's Day CEEFAX runs whole sections devoted to greetings sent in by viewers. Postcards are placed in a box, and several hundred lucky people will see their names in lights on their home television screens. In fact, the messages are so friendly, and some of the Valentine rhymes so original, that these pages become an entertainment in themselves, whether you are looking for your own message or not.

CEEFAX OPERATING PATTERNS

The CEEFAX pages are continuously updated for 17 hours a day — from 7 a.m. until midnight— seven days a week, 365 days a year. The CEEFAX newsroom is much like any other radio or television newsroom, although in operating practice it is perhaps nearer to a newsroom in one of the wire services, such as Reuters or Associated Press, which likewise have no specific deadlines and aim to keep the news flowing as fast as possible.

Sources of News

The main CEEFAX newsroom is at the BBC's Television Centre at the White City, in west London, some five miles from the center of town. Here, the journalists use news flowing in from the main international wire services and Britain's own Press Association, as well as from two agencies that are exclusive to the BBC: the General News Service (GNS) and the BBC Monitoring Service. GNS is very like an internal BBC wire service, drawing for its information on the other external wire services, but also from the more than 30 BBC newsrooms scattered about the country — from London to Belfast, and from Plymouth to Aberdeen. The BBC Monitoring Service, situated in a country house at Reading, keeps a listening watch on all the radio stations of the world, and relays all news of significance heard in English or any of the vernacular languages of the country doing the broadcasting. Quite often, the first indication of trouble in a foreign country comes via the BBC Monitoring Service, when it reports that a particular station has switched to martial music, is promising an important announcement in two hours time, or has simply gone off the air. Some of the most exciting BBC news scoops have come from this source.

One of CEEFAX's greatest strengths lies in being an integral part of the British Broadcasting Corporation, with its two national television networks, four national radio channels and more than two dozen local radio stations. All these are part of the same organization, administratively and operationally. CEEFAX also calls on all the BBC's specialized units and correspondents. These range from the motoring unit, consumer unit and farming unit, to the individual correspondents on Parliament, education and race relations. CEEFAX can make use of BBC reporters all over the country and the two dozen or so overseas correspondents in the main capitals of the world. The CEEFAX finance unit, led by two specialist editors, works closely with the BBC's main finance unit from an outpost at Broadcasting House, some five miles from the main newsroom. There it can draw on the financial expertise of other BBC correspondents and financial services.

The Newsroom Staff

At Television Centre the main CEEFAX news team, including the sports specialists, works a rather complicated shift pattern to cover the daily 17 hours. The team is led at all times by a person with the title chief sub-editor, but the number of assistants working with him will depend on the day and the time of day. Thus, the sports team will be strengthened on Saturdays and weak on Sundays, while the main news team will be stronger during daylight hours and when Parliament is sitting than it will be late at night. The chief sub-editor is responsible for copy editing, for briefing and supervising stories written by the other journalists, and for the general shape of the daily output. He will probably do the news flashes himself, type the headlines and thus set the pattern of priorities, and perhaps handle the major or most complex story personally.

As the chief sub-editor tends to become immersed in the detail of the news, CEEFAX has recently appointed a duty editor (working Monday to Friday) to exercise overall supervision. The duty editor, able to sit back one remove from the clatter of teleprinters and the busy hum of visual display units, can look at the output as a whole.

The CEEFAX newsroom. Chief sub-editors Will Garforth (foreground) and Ian Morton Smith are entering stories by means of a visual display unit. Research assistant Lys Holdoway is checking a story. After checking, the information is stored in the CEEFAX computer, and special equipment inserts it into unused lines of the regular television picture, for transmission over the air. © BBC, reproduced with permission.

The duty editor also concentrates on organizing and planning coverage for all future major occasions that are known about in advance, such as party political conferences and referendums, the World Cup and Wimbledon.

Preparing a Page

It takes a journalist a very few minutes to write the necessary 50 to 80 words for a CEEFAX story, especially if he or she is working from an existing script provided by the main radio newsroom and distributed by GNS. He will type directly into his visual display unit, which is an intelligent programmed terminal, making use of color to produce visually attractive "pages" which he can see in advance as they will appear on the television set. Ordinary letters and numerals are typed and colored by using hidden "control characters."

The journalist can also design bold captions and headlines, and a graph, logo or map to further enhance the layout and reinforce the information given in words. The CEEFAX system allows quite sophisticated drawings, albeit in computer-like shapes, and can even include cartoon characters, working diagrams and flashing indicators to tell the story more fully.

Once the journalist is happy with the page, and has proofread it for errors, he types in a coded instruction, presses the transmit button, and the page is entered into the computer for broadcasting. If it is a news story, for immediate use, it is fed instantly into the cycle of pages being transmitted — at the rate of four pages per second, as of early 1979. If the item is for "storage" and later use, he can instruct the computer to tuck it away for the following day's magazine, or for automatic release at a specified time to coincide with a particular event. Pages in quantity can be prepared in advance for special occasions such as a Budget Day or Christmas, or against an anticipated obituary or major news development.

Maps can be stored in the computer for immediate recall, to be reactivated and inserted at once to remind viewers, for instance, which exactly are the front line states in confrontation with Rhodesia or which countries border on Vietnam. The immediate provision of even a simple outline map, with perhaps a flashing arrow or cross to mark the place where a plane has crashed, or a hurricane struck, certainly enhances the basic teletext service.

CEEFAX has no resident engineers, for once started the system runs by itself, with only a minimum of maintenance and repair work needed. The little that is necessary is undertaken by the Television Network Department, the engineers who run the BBC's television operations.

COSTS OF TELETEXT AND COMPARISON WITH VIEWDATA

What are the major differences between the two systems in Britain, apart from the already-mentioned one of capacity — the simple several hundred pages of broadcast teletext versus the several hundred thousand of on-line viewdata?

Cost Differences

The flip answer is that teletext reaches the user for free. Of course, that is not really true; nothing in this world is free. But it is true that the cost to the broadcaster of establishing a countrywide teletext system is minimal, and this can be reflected when it comes to charging for the services. As previously mentioned, the BBC spent less than £200,000 to put CEEFAX on the air. Teletext needs no extra bandwidths, no extra transmitters, no extra power. The CEEFAX signals are simply squeezed onto just four out of every 625 lines of the ordinary television picture. They travel, as a series of almost Morse Code dots and dashes, at the very top of the television picture, unseen unless you have a very badly adjusted set.

Even the annual operating costs — mainly the salaries of the 20 journalists involved in the operation, plus certain payments to the wire services and other information providers outside the BBC —are still little more than petty cash in broadcasting budgets of nearly £300 million a year. In fact, direct costs of CEEFAX come to about £200,000 annually.

The costs of the BBC operation are currently funded out of the main Broadcast Receiving License that every set owner must have, and that in 1979 cost £25 per year for color and £10 for black and white. The CEEFAX charge against these sums represents only a few pennies per viewer. In the case of the commercial teletext service, the costs will presumably be met by advertising or sponsorship, and the first experiments in this direction were underway in early 1979. (ORACLE was off the air for much of 1978 as a result of an industrial dispute. In 1979 it was on the air five days a week, Tuesday through Saturday.)

So, the philosophy goes at the time of this writing, once the home viewer has bought (or in Britain generally, rented) his special teletext-equipped set, he receives three separate teletext services at no further cost. These are BBC-1 CEEFAX on the BBC's main network, BBC-2 CEEFAX on the second network and ORACLE on the commercial network of independent television (ITV). The commercial teletext service is exactly the same as the BBC service, though the two are rivals in programming. Which service you get will depend on which network you are watching when you switch from picture to teletext.

Renting a teletext-equipped set in 1979 cost £5 per month more than renting a color set without teletext. Purchasing the color set equipped with teletext outright set the consumer back about £1000; the teletext adaptor itself was £200, plus the cost of connecting it to an existing set.

In contrast to the absence of usage charges for CEEFAX and ORACLE, the Post Office service, Prestel, must charge for all its pages. The charge is either the simple cost of a telephone call, where an advertiser bears the input and distribution costs; or in most cases specific charges on a page-by-page basis, depending on the value of the information. These charges range anywhere from half a penny for most ordinary pages to the equivalent of a dollar for pages of a very specialized nature.

Size of Audience

Another major difference lies in the numbers who can use the services at any given moment. It should be remembered that in Britain in early 1979 only some 60% of homes had a private telephone, whereas 97% of homes had one or more television sets.

It makes no difference to the broadcaster (or to the electricity generating authorities) whether 20 people switch over to teletext or 20 million. But if all the 13 million people in Britain who indulge in a weekly gamble on the football results should seek to check their coupons on Prestel at the same moment on a Saturday afternoon, they would tie up every telephone exchange in the country, or simply not get through. The Post Office calculates roughly on a maximum loading at any moment of some 1% of their subscribers, but a broadcaster can accept 100% of the country's population at the same time without any effect on transmission.

Thus broadcast teletext, in terms of numbers, will always deal with mass audiences, while on-line viewdata will cater to specialized audiences (ranging from a few hundred to tens of thousands). The viewdata services will appeal to people actively seeking information at a particular time. The service will be most effective and productive the better it identifies and caters to a particular audience: say, commuters in the early morning, business people and house-wives by day, the racing fraternity in the afternoon and children seeking help with their homework, or the inspiration of a TV game or puzzle, in the evenings.

Hours of Use

One problem does emerge. If the Post Office aim is to extend the use of the telephone, particularly at times of day when the lines are under-utilized, there could be a conflict if the concentration is too heavily on business users. These will want most of their information at peak times, from 9:30 a.m. to 5:30 p.m., instead of the late evenings or after midnight when there is plenty of spare capacity. On the other hand, Prestel has the advantage of being available all the time — 24 hours a day — whereas broadcast teletext is only available when there is a television picture. This picture need not be anything more than a test pattern, on which teletext data can travel just as easily as it can on "Starsky and Hutch" or on a football game. In practice, as far as the BBC is concerned, this means that teletext is available for some 19 or 20 hours a day, from the moment that the Open University educational programs begin at 6:40 a.m. until the end of the midnight movie or the last talk show. It should be noted that, for countries that broadcast only an evening service, teletext could be a way of maintaining a broadcasting presence for a much longer span, using just that test card signal.

Waiting Time

Access to Prestel information is virtually instantaneous, once that first telephone connection to the computer has been made, whereas there is a definite waiting time on teletext. It is only measured in seconds, and depends on the number of pages in the magazine. As previously noted, BBC teletext pages in early 1979 were transmitted in sequence at the rate of a page every quarter of a second, or four pages a second. That means it took 25 seconds for the full cycle of a 100-page magazine, or a maximum wait of 25 seconds, a minimum wait of ¼ second, or an average of 12½ seconds. Twelve seconds may seem a long time, and often is when first encountered at a demonstration, with the viewer suffering from trade-fair feet and exhibition exhaustion; but in the comfort of one's home it is a negligible delay.

This waiting period, moreover, occurred with only a single-page memory in the decoder. Later versions were being built with four-page memories, and one enthusiast used a microprocessor to arrange his set so that it stored all 100 pages. A four-page memory will allow a viewer to store four chosen pages in his own sequence, and just click one after the other onto his television screen at his own reading speed. New ways of programming the BBC's second generation computer and transmission equipment are also designed to reduce waiting time. The new programming will allow many more pages to run as series or sequences,* permitting stories to be dealt with in greater depth and enabling a viewer to sit back and follow a whole family of pages without further action on his part.

Summary of Differences

In summary, the main points of difference between teletext and Prestel are easily stated. Broadcast teletext should satisfy those who want relatively small quantities of frequently changing information: news, weather, travel, share-prices, sports results. It should meet the needs of those who simply want a quick update on a situation, or a brief browse through the magazine. For those who want more specialized information for a particular purpose, and want it treated in depth, Prestel is their service. Instead of a brief summary of a company report or a rundown on the movements of overseas markets, they will be able to read a full summary of the chairman's speech, and compare past and present performances of that company, possibly illustrated in full graphic detail. Those who want

---

*This refers to the ability to send different (e.g., consecutive) versions of a given page on different cycles. It takes 25 seconds to send all 100 pages; this constitutes one cycle. If page 100 consists of an important news story, it could be continued on pages 100/2 and 100/3. Thus, 25 seconds later the second version of page 100 would be transmitted. A viewer who had page 100 on the screen would see it for 25 seconds and it would then be replaced by the continuation page. The new CEEFAX computer has the capacity for up to 100 "sub-pages."

the reasons why gold is weak and soybeans expensive, with a full analysis of comparative positions in a number of markets, will likewise find Prestel the appropriate service.

Users will have to pay for this information of course, but access to the pages of their choice will be immediate — at least, as long as their teenage children are not using the only telephone. This quip, of course, does make another point. The more traffic a telephone is required to carry the more likely the need for that second telephone in the home— not to mention the need for additional TV sets. Hence, both the Post Office telephone department and the industrial providers of second (or third) television sets for a family can look with benign countenance on the spread of both kinds of complementary videotext systems.

CONSUMER ACCEPTANCE

Britain has a head start on the rest of the world in the introduction of videotext services, but this is not to overstate the degree of customer acceptance of these intriguing and entirely novel devices.

For some, teletext was a slow starter, with only about 30,000 sets in the country after four years of operation (a guess was that two thirds of these were in private homes, one third in public places). It should be remembered that 1978 was the first year in which any sort of production line was in force. Prestel, too, was held back from its proposed pilot trial launch by union disputes over installation and a shortage of sets. In the end, the Post Office scrapped the pilot idea and went ahead with a public service from the start.

The adoption of teletext in Britain depends not only on the BBC's success in creating the right information, but on the efforts of the set manufacturers to convince customers to buy the gear. By spring of 1979 nine manufacturers were offering teletext-equipped sets: Rank Radio International, Thorn Consumer Electronics, Philips (Video Division), Kirby Lester Electronics, GEC Radio & Television Ltd., Decca Radio & TV Ltd., ITT Consumer Products (UK) Ltd., Pye Ltd. and Sony (UK) Ltd. At least one firm, Labgear Ltd., was making an adaptor to be attached to an existing set, while another company, Radiofin, had announced such an adaptor. First suppliers for all microprocessor components were Texas Instruments and Philips.

The estimate of the British Radio Equipment Manufacturers

The key to development of a mass market for teletext is a microprocessor that converts certain lines of the TV signal into words and graphics on the home screen. This is the first TIFAX decoder, made by Texas Instruments, Bedford, England. It measures 6 inches by 4 inches by ½-inch high. Courtesy BBC.

Association was that 50,000 to 100,000 sets would be installed by the end of 1979, and its collective sales goal was 300,000 sets by the end of 1980.

The speed of development depends largely on marketing. In the case of teletext, it is the rental companies (like Radio Rentals, Granada, Telefusion and D.E.R.) that seem most enthusiastic. Every color set they replace with a color-plus-teletext-plus-remote-control set is a chance to increase rental revenues. Buying a new color television, with teletext and remote control added, almost doubles the price of a television set. As previously mentioned, an adaptor to convert an existing set to receive teletext costs between £200 and £300. But for the many who rent in Britain, the *extra* to add teletext and remote control is about £4 to £5 a month on top of the ordinary rental.

Early buyers turned out to be as varied and across-the-whole-spectrum as the owners of television sets. Judging by the immense and immediate feedback reaching CEEFAX and ORACLE, viewers ranged from members of Parliament and airline pilots to dog breeders and the staff of foreign embassies. (Still, a survey in September 1978 showed that only 12% of the population were aware of CEEFAX.) Other immediate customers were pubs, clubs and hotels, for teletext is an ideal form of communication in public places. It is silent, and you can absorb the information in a bar or a hotel reception area without disturbing anyone else's pleasure or leisure. Prestel planned to introduce coin-operated machines, so that you could dial for information from a public library or hotel lobby and "pay as you go."

I believe that teletext will lead to a revolution in British social habits. The traditionally reserved Briton, in pub or club, will fall into conversation with whoever is operating or reading the teletext. The conversation may never rise above the British perennial, the weather, or perhaps who won the 3:30 p.m. race at Ascot, or the latest cricket score, but bar neighbor will talk to bar neighbor. I confidently predict that by the 1980s CEEFAX will provide the agenda for pub conversation.

TELETEXT ABROAD

The export of broadcast teletext has been conducted in a slightly lower key than the Post Office's ambitious marketing of viewdata.

Nonetheless, broadcasters and media persons from more than 103 different countries had visited the CEEFAX headquarters in London by the end of 1978. Without any real commercial involvement in the promotion of teletext (most of the BBC patents are protective, rather than commercially valuable), the BBC interest is more in standardization than in profit.

Sweden was the first European country to begin its own teletext trials, using the BBC's original experimental equipment. BBC engineers had previously carried out tests in both that country and West Germany. By spring 1979, two other Scandinavian countries — Finland and Denmark — had also undertaken evaluation tests. Despite the early start in Scandinavia, it seemed in early 1979 as though Australia might be the first country outside Britain to establish a public service. Engineers from the Australian commercial television companies were especially quick off the mark, and seemed to appreciate right from the start how valuable teletext could be in a country with particular communication problems because of geography and population dispersal.

Other countries actively involved by 1979 included the Netherlands, but interest had come from countries as far apart as Singapore and Mexico, Spain and New Zealand. Not all countries will follow the British pattern. (France has developed its own versions of teletext and viewdata.) But it does seem likely that microprocessors will be able to make almost any teletext and viewdata compatible, although it may involve extra cost to the viewer.

For countries using the same TV transmission standard as Britain, a complete teletext transmission system has now been designed as a package by a British software house, LOGICA. It is the LOGICA system, built to BBC specifications and under BBC engineering leadership, that has been installed at Television Centre to replace Esmeralda.

The package involves three DEC (Digital Equipment Corporation) PDP 11/34 computers, plus other specially designed equipment. Even for broadcasters who do not want to incorporate all the existing options, the price remains exceedingly small in broadcasting terms — between £120,000 and £250,000, depending on the sophistication of the system and the amount of backup required.

Although final specifications have not been written for broad-

A teletext system can display text in another alphabet, using a special read only memory (ROM) device. The illustration above shows a hypothetical table of gold, silver and bronze medal winners for the 1980 Olympics in Moscow, in Cyrillic characters. The Soviet Union is shown in first place. Courtesy BBC.

casting teletext on the NTSC system (the broadcast standard used in North America and Japan), experiments have been carried out in several countries.

The current state of teletext and viewdata in countries around the world is discussed further in Chapters 5 and 6.

FUTURE OF TELETEXT

It would be rash to make firm predictions about the future of teletext and viewdata in this age of developing electronics, micro-processors and computers. We are in the very early days yet, with teletext at the bicycle stage, and viewdata comparable to the first internal combustion engines. There are still the equivalents to come of aircraft and hovercraft, of jet engines and moon rockets, of space laboratories and Venus probes. There is no doubt that within a few years it will be easier and cheaper for television manufacturers to install teletext decoders as a matter of course. They are already no bigger in size than a large cigarette pack, and competition (as with the pocket calculator) is bringing the price down by leaps and bounds.

Effect on Newspapers

The dimensions of teletext and viewdata do keep changing, even if the technical specifications have now been set. At first it was thought that teletext would not involve hard-copy printouts, where-as viewdata eventually would. But within months, ITT had developed a printer for use with both teletext and viewdata, involving burned aluminum foil printouts, and a short time later a BBC engineer produced an even simpler version, using ordinary paper. The BBC was discussing leasing production of the printer with several companies in 1979, though no contract had been signed and no projected retail price was available. This hard-copy dimen-sion must be taken into consideration, even if its first applications are as straightforward as banks using it to circulate the latest exchange rates to tellers, gamblers to carry a list of tips to their bookies, or cooks to copy and preserve a recipe.

Even with a printer to provide hard copy, however, teletext will not mean the death of newspapers; its capacity is too limited. Viewdata will certainly make papers change some of their habits, perhaps even to the extent of making them see themselves first and

Teletext pages can be printed out as well as viewed on the screen. This is the BBC's hard-copy printer, still in a prototype version. Courtesy BBC.

foremost as information providers, with the means by which they provide that information being incidental. In Britain, at least, the further growth of CEEFAX and ORACLE will mean new relationships with wire services and other suppliers of information. Wire services and newspapers have considered CEEFAX as no threat because of the small number of homes it was reaching in 1979. As subscribers grow, news services might logically demand ever-larger fees for the information they supply, and newspapers might begin to complain about the competition. However, no single wire service enjoys a monopoly on news, and the BBC has virtually its own news agency within Britain, as well as a network of overseas correspondents and contacts.

Viewer Power

What is happening, with the arrival of teletext and viewdata, is a change in the viewer's relationship with his or her television set. No longer is the set a simple purveyor of such programs and films as the broadcaster in his wisdom thinks the public (or advertiser) wants at one particular moment in time. It is becoming something much more. It is becoming a home entertainment machine. It will display news and information via teletext and viewdata at any time of the viewer's choosing. It can be fed programs of one's own choice on disk or cassette. It can provide games and other entertainments. It is at last approaching the often-forecast role of being a home computer. Above all, it is a set controlled by the viewer, not by the broadcaster and/or advertiser. In the case of the teletext owner, he has the power to choose what to read, when to read it, and in what order to read it.

Educational Potential

An additional feature is television's increased potential for education. Not only do teletext and viewdata involve a return to words — to reading — but the medium is particularly valid for the kind of word games that teach children, or people learning English as a second language, to master advanced vocabulary.

CEEFAX has experimented with word games, quizzes, puzzles and even jokes based on word play, with immense success. Two separate educational experiments have been carried out — the first with 19 schools and technical colleges, the second directed at

```
279   CEEFAX 279  Mon 24 Jul  17:20/53

    WORLD CHESS
    CHAMPIONSHIP          2/3
    GAME THREE (Karpov-black)

    White    Black        16 B-B2   N-K2
 1  P-QB4    N-KB3        17 QR-K1  P-QN3
 2  P-Q4     P-K3         18 R-B3   R-K1
 3  N-QB3    B-N5         19 R-K3   B-B3
 4  P-K3     P-QB4        20 PxP    QxP
 5  N-K2     PxP          21 P-KN4  Q-B2
 6  PxP      P-Q4         22 P-B5   KPxP
 7  P-B5     N-K5         23 PxP    Q-Q3
 8  B-Q2     NxB          24 R-R3   NxP
 9  QxN      P-QR4        25 BxN    PxB
10  P-QR3    BxN          26 R-N1   K-R1
11  NxB      B-Q2         27 R-R6   R-K3
12  B-Q3     P-R5         28 RxR    QxR
13  Castle kingside       29 Q-N5   Q-N3
14  P-KB4    P-KN3        30 Q-R4   Q-K3
15  K-R1     K-B3         31 Q-N5   DRAW
Algebraic notation follows...
```

Teletext's ability to reach groups with special interests is seen in this display of moves of a world chess championship game between Karpov and Korchnoi. A viewer could "hold" his page throughout the match, and new moves would be entered as received. Courtesy BBC.

primary school students. The experiments were joint ones involving both the BBC's CEEFAX service and the commercial network's ORACLE. One significant aspect was the participation of two schools for the deaf. Because teletext is read, deaf people were not at the disadvantage they experienced watching TV programs with sound. It was also evident that, in addition to the specific educational projects covered in these two experiments, both primary and secondary level pupils were using teletext to keep up with current affairs. The response to questionnaires proved that this side of teletext interested the teachers almost as much as its potential for more traditional educational items.

BBC CEEFAX is the product of a public corporation, whose Royal Charter enjoins it to educate, inform and entertain. I hope we shall continue to do just that. CEEFAX has the ability to provide socially useful information in exactly the way the public wants it, and to display it all to best advantage in an electronic age. This is the real challenge of teletext services as we move toward the 21st century.

4

Viewdata: The Prestel System

by Max Wilkinson

INTRODUCTION

Viewdata is a simple idea that could have spectacular conse-
quences for business, employment and eventually the whole of
society. It was conceived from the marriage of two familiar pieces of
domestic equipment, the television and the telephone, to produce
something quite alien to most households: a computer terminal.
The television set in a viewdata system is modified with the addition
of some extra electronics to enable it to display text and graphics in
seven colors. The set is also connected to the ordinary telephone
handset, which can be used to dial up a central computer containing
a large store of information. This information may be of any sort,
ranging from airline timetables and encyclopedia pages, to business
statistics or theater guides. Indeed almost anything that is at present
published in books, magazines or newspapers could, in theory, be
stored and transmitted by a viewdata system. This new electronic
form of publishing is not going to replace the printed word, since in
its present form it is neither suitable nor economic for publishing
long articles or books.

Advantages

However, viewdata has one very important advantage over all
other forms of publishing, whether on parchment, paper or using
the newer materials like microfilm, magnetic tape or even video
cassettes. The distinguishing characteristic of viewdata is that it is
interactive. That means that the user does not merely receive
information as from a book, a film or a radio broadcast; he or she
can use the viewdata system to send information back to the central

Prestel display on home TV set. User is operating a simple keypad to select a desired page, in this case a cooking recipe. Courtesy British Post Office.

computer and can conduct simple conversations with the system.

In the pioneering system developed by the British Post Office, called Prestel, the user communicates with the computer by means of a small numerical keypad, rather like a calculator. This keypad is used mainly for telling the computer which "pages" of information the user wishes to inspect. However, it can also be used for making simple statements like "yes" or "no," for playing games with the computer, and for giving the computer numerical data in response to its questions. For example, Prestel can already be employed to help people with tax problems, mortgage calculations or legal advice. The system can also be used to make direct credit card purchases in response to advertisements stored in the computer. In Britain, users will shortly be able to punch in their credit card number to order a product delivered by mail. Alternatively, users could press one of the numbers on their keypad to indicate that they would like to receive further literature or catalogs from an advertiser.

The other great advantage of viewdata as a publishing medium is that information stored in electronic form can be frequently changed and updated. An airline company, for example, can alter the flight schedules stored in the viewdata computer every day if necessary. It can even show how many standby seats are available on each of its flights. This information is already provided by Pan American Airways and other airlines on Prestel's pilot service. The system has obvious advantages over newspapers for showing stock market quotations, sports results and many similar categories of information where speed and topicality are important.

Potential Uses

These possibilities are either already available on Prestel or they will be very shortly. Prestel is still in its start-up phase in the U.K., but became a limited public service during 1979. In the first year or two it will be used mainly by the business community for up-to-the-minute financial information. However, the British Post Office has ambitious plans to extend Prestel as rapidly as possible into ordinary domestic use. The Post Office has announced an investment of $46 million (£23 million) in 1979 and is prepared to spend up to $200 million on Prestel during the next five years to establish a network of computers throughout the country. It is aiming for 1 million users, with perhaps 1 million screen-sized "pages" of information stored in the system by the mid-1980s.

In principle, there is no limit to the size of a viewdata system and no obstacle to having a number of different systems operating in any country. Provided that all the systems store their information in the same format, a user could dial up any viewdata system that a publisher wished to set up. As viewdata systems develop, the interactive possibilities are likely to be considerably extended. The most obvious step would be to replace the simple keypad by a full alphanumeric (typewriter-like) keyboard. This would greatly extend the possibilities of talking back to the computer and of using the system to send messages to other users. A message-switching capability has already been demonstrated by the British Post Office, though there is some doubt as to the economic viability of using viewdata computers to send private messages.

An even more intriguing possibility for the future is the establishment of "intelligent terminals" in the home that could be used in national viewdata networks. The addition of viewdata electronics is really only the first step. It converts the TV set into what the computer industry rather slightingly calls a "dumb terminal," that is, an instrument that can receive and display information from a computer, but cannot do any processing on its own. The next step is terminals with substantial computing capabilities.

In the last few years, the development of microcomputers, etched onto a single chip of silicon and costing only $10 or $20, has opened the possibility of turning the television set into a small computer. A flourishing industry for hobbyists already exists in the U.S., and it is only a matter of time before television sets incorporating microcomputers start to be mass produced. These sets will be able to do double duty as home computers and as viewdata terminals, thus considerably extending the possibilities of viewdata itself.

A user with an intelligent terminal would be able to call up information from the viewdata library in order to perform his own processing. For example, a travel agent might use the viewdata files to obtain up-to-date information about flight schedules, and then perform the calculations needed to find the best route and the cheapest flights for the customer.

Similarly, a farmer might use the viewdata network to obtain the latest market prices for cattle and feedstock, then apply this data in his own accounting calculations. Another possibility, which has been demonstrated on Prestel, is to use the viewdata computers to store not information, but programs. A user with an intelligent terminal could rent a program from the library for a special purpose, and so greatly improve the power of his small computer. A viewdata computer program library is likely to be of most interest to small businesses, at least in the early years of the system. However, if intelligent television sets become priced low enough (as they probably will), a fantastic range of computer services could rapidly become available to the ordinary consumer. Thus, programs for medical diagnosis, legal advice and a wide range of educational and leisure purposes might be reproduced or adapted for home use. Once a market was created for this type of communication, it can hardly be doubted that enterprising publishers would start to develop new ranges of computer software products suitable for the ordinary consumer.

THE PRESTEL SERVICE

The actual operation of a viewdata system will depend upon the policies of the organization providing the service. However, the main outlines of what viewdata can provide and how it works are already being defined by the British Post Office, which invented the idea and is now actively selling it to telephone authorities and large companies elsewhere in the world.

The Post Office decided at an early stage to develop the system in partnership with two other groups, the television set makers and the organizations it calls "information providers." The TV manufacturers were crucial because the Post Office recognized that the success or failure of the venture would depend on stimulating mass production and mass marketing of sets capable of receiving the new service. As for the information in the viewdata service, the Post Office rejected the model of the British Broadcasting Corporation (BBC), which controls all the programs broadcast by its transmitters. The Post Office decided to adopt the opposite policy of exercising no say whatever over the information provided by the Prestel service, except what was required by public decency and the law.

The Post Office regards Prestel as analagous to the telephone service — a way of conveying whatever information individuals desire. It therefore sought out different organizations that might be interested in providing information for the new service. This policy has been so successful that there is now a waiting list of organizations that wish to become information providers by leasing pages on the Prestel computer. So far, more than 150 publishing and other organizations have contracted with the Post Office to supply nearly 200,000 pages of information under a wide variety of headings. Each of these pages represents the amount of text or graphics that can be displayed on the television screen at one time. This is a maximum of 24 lines of 40 characters each, or a total of 960 characters.

There were about 1100 users during the test service at the beginning of 1979. Each is allocated a personal code number that is built into the receiver; the number is automatically scanned by the Prestel computer every time a call is made, and is used for identification and billing purposes. To obtain Prestel information, the user switches on the set and makes a local telephone call to the nearest Prestel center. In some of the earliest sets it was necessary to dial all the digits on the telephone handset, but in most of the newer models, telephone connection is made automatically by pressing a button on the television set.

At another touch of a button, the Prestel index appears on the screen. This index, listing the main categories of information, leads to a series of more detailed indexes that will guide the user to the precise page he wants. Alternatively, a user may consult a printed directory, which should enable him to obtain the exact page he requires immediately.

Charges to Users

While connected to the Prestel computer, the user is incurring three separate types of charge. First, he is paying a local telephone call charge for the period of connection. Second, he is paying a time charge for the period during which he is logged onto the Prestel computer. This was set at two pence (four cents) a minute for the start of the service, but the Post Office says the charge may be varied later. Third, the user pays a charge levied on each frame by the information provider. This charge may be zero in the case of information provided by advertisers or public bodies, or up to perhaps 50 pence ($1) for information of particular commercial value.

Information providers write their Prestel pages by means of a special terminal, shown here. Courtesy British Post Office.

In early 1979 most of the high value information was priced at about 20 to 30 cents per page, while general information aimed at a wider public was priced at between one cent and six cents per page. The pricing structure is likely to vary considerably, however, as the pattern of usage begins to be clearer.

The Post Office insists that the price tag for each page should be displayed at the top right-hand corner of each page. In addition, the computer will give each user a running total of the amount spent during each Prestel session and the total billing for that quarter. The charge levied by the information providers is collected from the user by the Post Office, which then passes it on after deducting 5% to cover the costs of collection and bad debts.

Storage Facilities

The Prestel computers, located in telephone exchanges, are designed to hold 250,000 pages each and to handle up to 200 calls simultaneously. The number of pages can be expanded by the addition of more disk memory. The longest waiting time for any particular page is designed to be two seconds. A further four seconds is required for the page to be "written" on the user's television screen. However, the user can start to read the page as soon as the first line appears, so that this waiting time is not noticed in practice. Each computer is to serve its own group of users and all will carry identical information, which is kept up to date from a central bureau. In 1979, the Post Office had planned to spend $10 million (£5 million) to set up Prestel centers in four of the largest cities: London, Birmingham, Manchester and Edinburgh. (Service in the latter three cities was postponed until 1980, however.) A further $36 million (£18 million) was earmarked to extend the service to Cardiff, Glasgow, Leeds, Liverpool, Norwich, Nottingham and other important centers.

When the Prestel service is further developed it is expected that a new central storage facility will be added, called a "data warehouse." The warehouse will store pages that are infrequently inspected, and are therefore not suitable for duplicate storage in all the local computer centers. If a user wants to see a page stored in the warehouse, he will dial up the local computer in the normal way. The local computer will then obtain the page from the warehouse by a high-speed data link.

PRESTEL   Page 0a

1 LIST OF INFORMATION ON PRESTEL
  News & Weather   Sport & Hobbies
  Entertainment   Holidays & Travel
  Marketplace   Jobs & Careers
  Advice   Books & Reference
  House & Garden

2 ALPHABETICAL LIST OF CONTENTS
3 LOCAL INFORMATION
4 INTERNATIONAL INFORMATION
5 BUSINESS PRESTEL
6 SPECIAL DATABASES

7 GUIDE FOR PRESTEL USERS
8 GUIDE FOR INFORMATION PROVIDERS

Prestel ™

Main Prestel index. Various printed indexes enable the user to find a specific page more quickly. Courtesy British Post Office.

The cost of storage in a data warehouse is relatively much lower, but the cost of inspecting it is higher because long distance communication is needed. It is likely, therefore, that the warehouse will be used mainly for reference material, like electronic encyclopedias, which can be left unchanged in the system almost indefinitely and which are inspected relatively infrequently. However, the warehouse may prove especially suited to two other types of information: first, local or regional information, like theater times and local classified advertisements; and second, items like commodity prices, which are constantly being updated.

It may seem paradoxical that local information could better be stored centrally. Nevertheless, it may prove cheaper to do this than to duplicate local information in computer centers throughout the country. Although an earlier idea was that local computers might carry information relating only to the area they served, now it is felt that a user in one part of the country may wish to see, for example, hotel vacancies or timetables in another. For this purpose central storage would be more appropriate. The other use of the central warehouse — for frequently updated information — could encompass stock prices, availability of airline seats and, eventually, a national classified advertising service for automobiles and houses. In many such cases the cost of keeping files up to date may outweigh the costs of distributing pages to satellite computers. However, nobody yet knows precisely how the economics will work out, and much will depend on the relative fall in price, during the 1980s, of communications equipment and circuits vis-a-vis that of computer hardware.

It is clear, however, that by early 1979 the stage was set for a vigorous competition between rival information providers, who will have to develop a completely new range of skills and pricing policies to compete for the users' interest. So far no other viewdata service has been established in the U.K. in competition to Prestel, but the system may to some extent compete with the BBC's broadcast teletext system, called CEEFAX, and the comparable service from the independent television network, called ORACLE.

Teletext pages, which look exactly the same as Prestel pages, are broadcast on the television channels simultaneously with the television programs. Teletext is free of charge and does not require a telephone connection. It can be obtained, with a suitable decoder, directly off the air, which means that Prestel sets can display teletext information if the sets have the proper decoders built in. As described in Chapter 3, however, teletext is not interactive and is relatively limited in scope, since only a few hundred pages can be transmitted per channel.

ORIGINS OF VIEWDATA

The development of viewdata in Britain has been remarkably rapid. The establishment of limited public service in March 1979 was two years earlier than scheduled in the original timetable. This is because the Post Office decided to move directly from a year's test service to a public launching without the planned market trial.

The rapidity of viewdata's introduction has created some acute problems for information providers and set makers, who had originally expected to be helped by two years' market trial to establish rates of usage. Manufacturers had not produced enough sets to make possible extending Prestel to locations outside London in 1979, thus causing a delay in this part of the schedule.

However, the Post Office concluded at an early stage of the development that viewdata is essentially a mass market medium. Its economics, according to the Post Office, require a large number of users and a large spread of pages; the sooner this condition is reached, the better. As Alex Reid, head of Prestel, remarked: "If you are trying to achieve take-off, it is foolish to do it at half throttle." (It should be pointed out that critics of the Post Office approach feel that viewdata is not a mass public service, but initially a business service. Thus, they argue, large investments aimed at reaching many homes are ill advised.)

In any event, the service reached a critical commercial point only five years after the original specification and design study was completed by Sam Fedida, the Post Office research engineer who invented the idea.

Fedida started to develop the idea during the late 1960s, when he was working on a computer system to keep track of vacancies in European hotels. He found that 80% of the cost of this system represented salaries and overhead in the central bureau that operated the computer information service. This was because the system required clerks to interrogate the computer on behalf of the customer. How much better, Fedida thought, if the system could be designed so that the customer could look up available vacancies for himself. The system would become "self-service" and a large part of the labor costs would be saved. These ideas began to achieve a more tangible form when Fedida joined the Post Office as a researcher in 1970; there he came into contact with a number of people working on Viewphone (called Picturephone in the U.S.), the system that allows telephone callers to see, as well as hear, their interlocutors.

From experiments with adapting Viewphones to allow them to display data — words and figures only — the Post Office researchers had the imaginative idea of using the ordinary television set. At the time, television sets were firmly categorized as belonging to a world of entertainment —far removed from the serious endeavors of telephone and computer engineers. The technical problems of displaying data on a domestic television set are relatively simple. The conceptual leap required was to foresee how mass-produced computer terminals could be exploited in a public system.

By July 1974 Fedida and the small team of engineers he had gathered around him were ready to show a working model of their new toy to Sir Edward Fennessey, then deputy chairman of the Post Office and managing director of Telecommunications. Fennessey was enthusiastic, and immediately asked the marketing division to explore its implications. Of course, the fundamental lure of the system to the Post Office was its promise of increased telephone usage. Only about 60% of the households in Britain have phones, well below the rate in the U.S., Canada or Sweden. Use of the phone network after business hours is especially low. Thus, a service that encouraged information retrieval by customers in their homes would boost telephone usage and perhaps even increase phone installations.

While the engineers worked on the details of how to organize the information storage, the major principles for the system began to be formulated. They were:

1. It must be reliable. A system that frequently broke down would soon be discredited.
2. It must be simple enough for anyone to use without instruction.
3. It must be very fast, so that as many people as

     possible could be rapidly served by the computer.

4. The cost must be very low. Both the sets and the computer centers must be designed to be as inexpensive as possible.
5. A network of computers storing identical data, rather than an interconnected network different data bases, was to be established.

In spite of a number of quite serious problems during the development stage, in 1979, it appeared that these broad objectives were being achieved.

THE TECHNOLOGY

The basic equipment needed for viewdata is a minicomputer with specialized software to allow as many users as possible to inspect the stored information (data base). It must, above all, be reliable. In the U.K. the minicomputer used is made by General Electric Company (GEC - no relation to General Electric of the U.S.). It is a GEC 4080 with disk storage units capable of holding 70 megabytes of information each (a megabyte is roughly the equivalent of 160,000 English words).

The Prestel system is designed to have 200 "ports" on each computer (i.e., it will accept 200 simultaneous users) with a maximum waiting time of two seconds per frame.

Because of problems with software the goal of 200 ports had not been achieved by the early part of the test service at the beginning of 1979. However, the Post Office is hoping that continuous development will allow the number of ports to be expanded. This is very important for the economics of the system, because a larger number of ports will allow a greater usage of the stored information and, therefore, more revenues for both the Post Office and the information providers for a given capital outlay.

As previously noted, the format of the viewdata pages is 24 lines of 40 characters. Each character is defined by a 7 x 5 dot matrix. The colors available are red, green, yellow, blue, cyan, magenta and white. The system is connected to the telephone network by a simple jack and socket. Data are transmitted from the central computer at the relatively slow rate of 1200 bits (computer pulses) per second. Transmission of information from the user back to the computer is at the very much slower rate of 75 bits per second. This allows for the fact that the user will be sending only relatively simple information like page numbers and "yes/no" responses to questions. The character "envelope" for the encoding of data is 10 bits, of which seven are used to define each new item of data. That means that in one second 120 items of data can be sent from the computer or 7.5 items can be returned by the user.

When fully digital (computerized) telephone networks are introduced, higher grade lines will be needed with a much greater information-carrying capacity. In the U.K., digital telephony is expected to use transmission rates of 64,000 bits per second, or roughly 50 times the rate of the present Prestel system. Even using the present telephone network, it might be possible to increase the rate fourfold to 4800 bits per second. This would be fast enough to transmit black and white pictures or color pictures in rather crude tones. The question of how these improvements will be phased into the system was still under discussion in spring 1979. One compromise, for example, might be to introduce small pictures occupying only part of the screen.

The quality of reproduction depends not only on the transmission rate on the telephone lines, but also on the capacity of the electronic memory added to the receiver. After transmission, each page of information is stored in the set and can be viewed indefinitely, even after telephone connection with the computer has been severed. The higher the definition of picture or graphics, the greater the memory required and therefore the more expensive the set will become.

Prestel's graphics are based on dividing the screen into 6000 individual cells. The Canadian Telidon system divides the screen more finely, into tens of thousands of cells. A high definition color picture would require a cell structure of a different order of complexity, with correspondingly more memory. As the system develops, therefore, a series of interesting decisions will have to be made about what level of technical complexity the market can sustain.

COSTS AND PROFITS

In terms of its storage capacity, the system is expandable to almost any conceivable size. As many disk units as necessary could be added to the computers — at least up to the capacity of several million pages. However, from a practical point of view, the expansion of page storage will have to be related to the number of users. Clearly if a very small number of users is connected to a very large number of pages, each page will be inspected (on average) a small number of times each year. Consequently the charge per page would need to be very high. The rate at which pages can be inspected is also limited by the number of ports offered by each computer, and its speed. The following calculation based on the Prestel system illustrates the point.

Suppose the number of ports per computer is 200. Suppose the system is used for an average of eight hours a day at full capacity with each user hooked up to the computer for an average of 30 seconds for each page inspected. With the system working flat out, that means two pages could be inspected each minute through each of the 200 ports. That is 400 pages a minute for each computer, or a maximum of 200,000 pages in eight hours' average use per day. In the Prestel system a large number of identical computers with identical data will be in use. Consequently, with five computers in the system the maximum number of pages which can be inspected per day is 5 x 200,000 = 1 million. With 10 computers, the number of pages inspected would increase to 2 million, and so on.

The calculation of how often, on average, a page can be inspected is clearly related to the costs to the user and the profit that can be expected by those providing information. Suppose, for example, that the average cost of maintaining a page of information in the system is $40 per year (including Post Office charges, the cost of obtaining and updating the information, salaries and overhead). That means that the average cost of maintaining a page would be about 11 cents per day ($40 ÷ 365).

Now suppose, for simplicity, that the system consists of 10 identical computers each with an identical store of 2 million pages. Following the assumptions above, it is evident that 2 million pages is the maximum number that can be inspected on any one day (through the 2000 ports in 10 computers). Therefore, in such a system, the providers of information can expect each page to be inspected once a day on average (though some pages would be used much more often than others). To break even, information providers would have to charge an average of 11 cents a page.

These figures, though based on the configuration and likely costs of Prestel, are illustrative only. They are meant to show that for any given assumptions about the usage of a system (and eight hours a day is probably too optimistic), the charges that must be levied to recover costs are directly related to the number of computers in operation and to the number of pages stored in them. If, for example, the number of computers were 10, but the number of pages stored were only 1 million instead of 2 million, each page could theoretically be inspected twice as often, so that the charge per page could on average be halved to 5.5 cents. Alternatively, if the data base were expanded to 4 million pages, the charge per page would have to be doubled to 22 cents.

The attractiveness of a viewdata system is therefore very much dependent on achieving as large a number of users as quickly as possible. The more users, the more computers needed to serve them and hence the more pages that can be stored at any given price level.

THE INFORMATION PROVIDERS

Since viewdata is essentially a publishing medium, anyone setting up a system must first decide who is going to use it and how they will do so. The British Post Office decided that it would run all the technical aspects of the service, but that any organization should be allowed to contribute information. It set charges, however, in such a way that only relatively large organizations could afford to become information providers, especially at a time when investment in the service was necessarily speculative.

Two classes of service were envisioned: the first, Class A, was intended to apply to pages that would be duplicated in all the regional computers. The second, Class B, was for pages that could be expected to be stored in the data warehouse.

The charges for the Class A public service, starting in 1979, were: £4000 ($8000) service charge plus an annual rental of £4 ($8) for each page of information stored in the system. Discounts of 25% and 40% were offered for contracts of three years and five years, respectively. For Class B, the service charge was only £1000 ($2000) plus an annual page rental of £1 ($2). In addition, however, the Post Office will levy a charge of 0.5 pence (one cent) for each time a user inspects the page. The lower annual charges reflect the relative cheapness of storage in a central warehouse, and the usage charge reflects the expected cost of communication from the warehouse to the satellite computers. Class B information providers may also be restricted in the times of day during which they may edit their pages.

As of early 1979, a wide range of organizations had signed up for pages in Prestel. For example, Fintel, a joint service of the *Financial Times* and the *Exhange Telegraph*, was making available stock prices for companies on the London Stock Exchange. It also provided summaries of economic conditions in various countries, summaries of brokerage house reports on specific stocks and other business statistics. ABC Travel Guides, a publisher of timetables, showed schedules for intercity rail and air travel. Norwich Union, a major insurance company, had pages explaining its various policies (these pages would be no charge to subscribers) and inviting users to call its various agents, whose phone numbers were listed. Pan Am and British Airways listed flight schedules and, in some cases, availabilities of seats, while MacDonald Educational, a book publisher, had signed up to put children's stories into Prestel. Another publisher, Mills and Allen, offered advice on health, budget travel and gardening, as well as games and puzzles, and Caxton Publications announced plans to condense a 20-volume encyclopedia into 10,000 frames of information. The Central Office of Information provided large amounts of government data.

Information providers need to rent or buy special editing terminals. To edit one of his pages, or to supply a new page, the provider calls the system through the telephone network and enters a special code number. He then calls up one of his allocated pages, which will be displayed on his television screen. He may change any part of the page, or if it is blank, enter new information altogether. When he is satisfied, he presses a combination of keys, which will place the modified page back in the computer files.

As of spring 1979, most editing was done while the information provider was connected ("on-line") to the Prestel computers. Eventually, however, more and more editing will be done "off-line," using the information providers' own computers or intelligent terminals. This will ease the burden on the telephone system and on the Prestel computers.

Although the techniques of entering information are likely to be similar in all viewdata systems, the structure of information providers may differ considerably. In the U.K., for example, a large diversity of providers has been encouraged, but in Germany it may be that the system will be dominated by a number of large publishers who will obtain information from secondary sources. Already in the U.K. some organizations have booked large blocks of pages which they are preparing to sublet to smaller companies that would prefer to use the system through an agency than to deal directly with the Post Office. Baric Computing Services, owned by International Computers Limited (ICL) and Barclays Bank, is an example of a company that has set itself up as an intermediary to put other people's information into an attractive format. In spring 1979 its pages included consumer guides, theater guides and games.

Market competition will play the main role in determining the future shape of the Prestel system, as information providers try to develop new ways of making their material attractive and valuable. They will have to make complex judgments about pricing versus the frequency of access expected, and they must, above all, develop efficient ways of indexing and cross-referencing, which will enable casual users to find their pages. Groups of information providers serving a similar clientele may well combine to provide a common index, or they may try to separate themselves as much as possible from their rivals.

Already in Prestel's early stages a number of intriguing signposts can be seen. In the first place, it is clear that publishers who have relied on selling official information in a convenient booklet will be up against stiff competition from the official sources themselves. However, publishers who combine and reorganize information from a number of governmental or commercial sources (e.g., publishers of airline timetables) would seem to have an important role in a viewdata system. Secondly, in Prestel as in print publishing, a company that can provide valuable information while avoiding the costs of frequent editing will make more money than a company that has to change its frames every day—unless the second company can charge a much higher price.

These issues will tend to be focused much more sharply in a viewdata system than in conventional publishing, because the user will be paying only for the information that he chooses to inspect. He will not be making a broad general comparison between one magazine and another, but between each item he reads and the price displayed at the top right-hand corner of the frame. He will be constantly invited to ask the question, "Was that worth it?"

This point will also be felt acutely by advertisers and public authorities that offer their pages free of charge. Viewers will be paying a telephone charge while connected to the system; moreover, they will have to make a conscious decision to dial up an advertisement page or a public announcement. Advertisers and public propagandists will therefore have to think of new ways to make their material informative and attractive. If advertisers succeed, the interactive possibilities of the system will open up new techniques in selling. Salesmen will be able to ask questions of their potential customers and, in the case of insurance companies, for example, lead the customer through personal and financial calculations. Direct sales can also be made to customers who enter their credit card numbers into the system.

CLOSED USER GROUPS AND PRIVATE SYSTEMS

Prestel was conceived as a public network in which most users could have access to most of the stored pages. However, the system will also accommodate closed groups of users who do not wish outsiders to gain access to specialized or confidential information. Companies, financial institutions, professional associations and trade unions, for example, may thus make use of the public information network for their private purposes. All that is required is a special key number or "password" that will allow access to the files and that will prevent outsiders from seeing the information.

Some companies may, perhaps, use the system for catalogs and price information that traveling salesmen could obtain from any Prestel set in the country. Or they may use it to communicate between different branch offices. In this kind of application, Prestel will operate in the same way as other computer-based information systems already in existence. The main difference is that it will be simpler and five to ten times cheaper. When a wide network of Prestel receivers is available, it may be possible for a salesman visiting a customer to check the availability and price of his company's goods on the spot and even order them directly through the system.

Private viewdata systems can operate in exactly the same way as the public system, except that the private system operator would need to buy his own computer. Anybody owning a Prestel receiver could dial up a compatible system, provided he knew the telephone

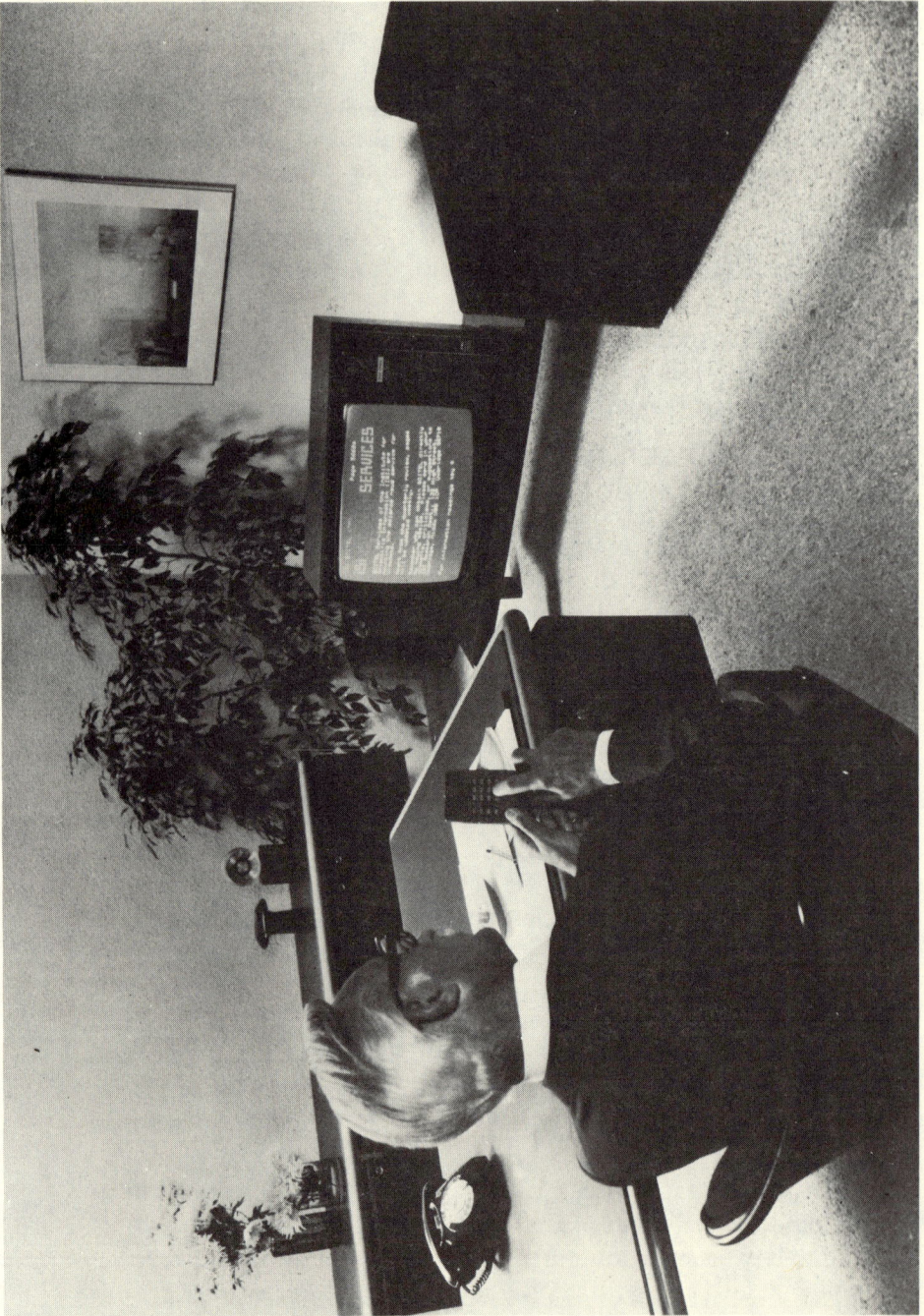

Business users of Prestel were expected to predominate at the outset. Courtesy British Post Office.

number and the correct code for gaining access to the system. However, there is one important limitation. In the interests of cheapness, the modem (which converts the computer pulses into a form suitable for transmission by telephone wires) has been designed to be suitable only for local calls. Because the Prestel system is designed around regional satellite computers, users will make only local calls. All other transmissions of Prestel pages will be made by high-speed data links between different computers.

A private operator who wanted to take advantage of the mass-produced Prestel receivers would therefore have to restrict his system to users within local telephone range, or he would have to set up his own network of computers. Completely private systems are therefore likely to be set up mainly for internal information within companies or government departments, although some commercial systems could become available in large cities like London. In theory, the Post Office could allow nationwide private systems with their own computers to be connected into the public Prestel network. However, because determination of rates would be complicated and confusing, Post Office opinion seems to be moving against this idea. In Germany, however, serious consideration is now being given to the establishment of a network of computers that would all carry different information (instead of identical information, as in Prestel). In such a federal system, the possibilities of linking in with private systems might be more attractive.

PRESTEL RECEIVERS: STANDARDS AND COSTS

Since low cost achieved through mass production is the essence of viewdata systems, great emphasis will be placed on efforts to make systems as similar as possible. However, the adaptors for television sets in different countries will have to conform to those countries' different regulations. The most important item is the modem, which must meet prescribed standards. In the U.K. the British Post Office has allowed set manufacturers to incorporate an approved modem (the device is etched on a single chip of silicon). In addition, special protection is required to prevent the high voltages of the color television set from accidentally being applied to the telephone network. The Prestel system's standard of data modulation conforms to the International Standard CCITT V23, and safety protection of the terminal must conform to British Standard BS415.

Prestel's ability to function as an electronic mail system for sending personally addressed messages is shown here. Courtesy British Post Office.

At the beginning of 1979 a color Prestel set cost about $2300. This was partly because a sophisticated set with remote control was needed, but mainly because the extra components were not yet mass produced.

It is expected that small black and white sets should be available at about $400 quite soon. By the mid 1980s the Post Office hopes Prestel sets may cost no more than about $100 to $150 in addition to the price of an ordinary TV receiver. It is possible also that the Post Office may develop adaptors to be fitted near the telephone, which would plug into the aerial socket of a standard television set. These would be rented to subscribers.

In November 1978 the Post Office identified the following companies as having met technical and safety requirements for Prestel sets: Cherry Leisure, Decca Radio and Television Ltd., GEC (Radio and Television) Ltd., ITT Consumer Products Ltd., Pye Ltd., Pye Labgear Ltd., Rank Radio International, Real Flex, Rediffusion Consumer Electronics Ltd. and Thorn Consumer Electronics Ltd. Also developing Prestel sets, adaptors or business terminals were Kirby Lester Electronics, Philips Electrical Ltd. and Standard Telephones and Cables Ltd.

INTELLIGENT TERMINALS AND TELESOFTWARE

Although viewdata is beginning as a straightforward information retrieval system, expanding from business to domestic use, it may in the longer term develop more of the characteristics of a computing network. In principle, most small computers or computer terminals can be hooked up to the system through suitable connecting equipment. It is likely, therefore, that the network will include many small business machines and home computers.

One of the features of most existing computer networks is that programs (software) stored in the central computer can be sent out for use by the smaller machines in the network. This could only be made to work in the viewdata system if a common programming language, acceptable to different families of microcomputers, could be designed. The British software company CAP-CPP has set out to do just this, and claims to have achieved it with a concept which it calls Telesoftware. The company says: "The present operation of the (Prestel) system only scratches the surface of its potential." Using a new language called MicroCOBOL and a special design for the viewdata system, the company claims to have developed the possibility of running the same program on nearly a dozen different types of microcomputers. It is also developing a similar version of another computer language, BASIC, for approval by the British Post Office.

Programs written in one of these "universal languages" could be stored on the viewdata system in the ordinary page frames and called up for use on any intelligent terminal or microcomputer, whatever its operating system. CAP-CPP believes that one of the main contributions of this development will be to take the mystery out of computing and to extend its capabilities much more widely than at present. A user who called up one of the programs in the telesoftware library would probably have little idea of how the program worked. He would simply be taking delivery of a routine with easy-to-follow instructions, which would appear on his screen to prompt him.

The main disadvantage of using viewdata systems to store and transmit software is their relative slowness. The standard Prestel frame can store 400 bytes (characters) of program. A typical program might consist of 12,000 bytes, which, at a rate of five frames a minute, would take six minutes to load from the Prestel system into the minicomputer. This is several orders of magnitude more slowly than can be achieved in a specially designed computer network. Furthermore, telesoftware operates perhaps 10 times more slowly than a program especially written for a particular machine.

It is clear, therefore, that telesoftware and viewdata will never supplant dedicated computing networks, nor are they intended to do so. The main point is that the efficiency is quite high enough for many applications in small business and the home. You don't usually need to travel by jet airplane if you are only going to the next town.

EXPORTATION OF VIEWDATA

Soon after the development of viewdata got under way, the British Post Office started to think about the possibilities of exporting it to other countries. There were several reasons for this. First, there was the understandable British pride in a new invention which seemed to open wide vistas of commercial exploitation. Then there was the hope of helping British manufacturers to export computers and television sets adapted for the system. But perhaps the most important motive for exporting was the desire to prevent a large number of incompatible systems from being developed in different parts of the world. Since the success of viewdata must be based on the availability of cheap mass-produced equipment, it was obviously desirable to achieve as much of a common standard throughout the world as possible.

In this aim the Post Office has had a mixed success. Its most important success was in selling the system to the West German Bundespost. Germany is the largest potential market for the system in Europe, and it uses the same PAL system for color television as the U.K. Britain, Germany and Holland (which has also bought the system) will therefore form a strong nucleus for a European market. Switzerland and several other European countries are likely to follow. Another country to buy the viewdata system was Hong Kong.

At the beginning of 1979 a strong effort was mounted by Insac, the British Government's computer software marketing company, to sell Prestel in the U.S. Elsewhere different systems are being developed by different countries, although most are loosely based on Prestel. Insac has licensed U.S. rights to Prestel to General Telephone and Electronics.

Rival systems under development in the United States and other countries are described in Chapters 5 and 6.

THE FUTURE OF VIEWDATA

The British consultants Butler Cox, who have undertaken a study for 37 major U.S. organizations about the prospects of viewdata in North America, say: "Our research has established quite clearly that viewdata systems are coming to the countries of the world in one way or another. The commercial pressure mobilized on their behalf will be too great to fail."

Alex Reid, head of the Prestel division of the British Post Office, also believes the service has enormous possibilities. In May 1978, when final plans for the test service were being laid, he said: "Our aim is to produce a cheap and universal means of electronic publishing, available to all; a new medium of communication, comparable in scale to radio, television or the press." Reid admits that, "As with any new product there is a risk that the customers will not take to it." But in addition to being encouraged by market research and other indicators he says: "We have, in any case, a gut conviction that this kind of service will come." He hopes that by the end of the 1980s Prestel's users will be numbered in millions, with millions of pages stored in the system. The system will then have economically "taken off." Costs will fall sharply, and users and information providers will mutually encourage each other to greater uses of the system.

The British Post Office believes that a self-perpetuating momentum will be achieved in two phases. First will come the saturation of the business market, for which the cost of receiving information will be much less important than the question of whether that information is genuinely useful. As noted previously, business users may be expected to pay 20 cents to 40 cents for an inspection of each page, perhaps 10 times the amount that will be paid by the domestic user. If this expectation proves correct, the system could be viable with several hundred thousand users in the U.K. with its total population of 56 million people.

One difficult question is the amount of Prestel usage required for the Post Office to earn a return on its enormous investment. At four cents (two pence) a minute for connection charges, .5 cents to three cents a minute for phone charges and zero to 20 cents a page for page charges, the average household might use Prestel for an hour a week. This would come to an expenditure of $300 (£150) a year, exclusive of equipment purchase or rental. A business user plugging into Prestel 30 minutes a day, with higher page charges and peak phone rates, might wind up paying $1300 to $2600 per year.

From the Post Office's point of view, its mandate from Parliament is a return on capital of 5%. Revenues for storing 450,000 pages from information providers would yield $4.6 million (£2.3 million) and average connect charges from 10,000 business users might provide $2.5 million. On an investment of $70 million (the projected investment by the end of 1979), that comes to 10%, before added phone revenues. However, after operating expenses and interest charges are deducted, the return on capital might well be reduced to 5% or less. With 100,000 users, revenues from connect charges alone could be $25 million (£12.5 million).

The Post Office figures the system can become a totally public medium when it reaches a take-off point of, perhaps, 1 million users in the U.K. This is the estimated figure needed to bring costs of receivers and of information down to a "consumer level."

If there is a danger in the Post Office projections, it is probably that of overestimating the early demand for viewdata systems and underestimating the prices that will have to be charged to recover the costs. Even attracting several hundred thousand business customers in the first year or two is no mean task. If the market should prove much smaller than this number, the prices charged for the information would have to be much higher.

Prices of 10 or 20 pence (20 or 40 cents) per page may seem high in a consumer context, but for business and professional uses they are low. Consider a standard business credit report, taking up no more than two printed pages, that would probably fit on two or three Prestel frames. Presumably, subscribers to a credit reporting service could pay an annual subscription fee and receive a special code number enabling them to have access to the reports on a viewdata system. In the U.S., such a report would cost $5 or $7, delivered by mail by Dun & Bradstreet. Would the customer balk at paying several dollars to have it on the TV screen instantaneously? Yet this modest price is five or ten times the charges being quoted for business information.

The method of charging in Prestel, i.e., by the page, is also open to question. Both in telephone systems and in computer time-sharing businesses, customers are accustomed to paying by the minute or hour of usage. Forcing the customer to constantly make a yes/no decision about whether to look at more information may impede use of the service. The U.S. licensor of viewdata, Insac, concluded that charging by the unit of time is the only way to market viewdata. The U.K. proponents might also come around to this point of view eventually.

A final, skeptical point relates to the level of investment announced for Prestel. Although the service's organizers talk about how inexpensive and efficient it is, compared to other computerized information systems, the level of investment being made to launch Prestel (at least $100 million) is anything but modest. It seems doubtful that a private company would ever plunge into the market in this way without a far more extended trial, and without limiting its risk.

Assuming public acceptance of the basic service is achieved, a new phase of technological advance will likely follow quickly. This will probably be centered on the use of computing devices in the home as well as in the office with, importantly, the addition of electronic printers and magnetic memory in the form of cassette recorders or mini disk units.

The addition of printers may prove most important, since they will remove the main disadvantage of a viewdata system — that it is ephemeral. Once a printer is added, the information on the screen can be printed out to be read at breakfast or on the train in the morning. Moreover, printers, especially if they are silent, could do their work at night during off-peak hours, when the system has plenty of spare capacity. When such machines are available at a reasonable cost, the era of electronic newspapers and electronic mail will have arrived.

Information similar to that in many newspapers could be transmitted overnight and automatically printed out during the quiet hours in thousands, and possibly in millions, of homes. But even if newspapers were not directly threatened, their revenues, particularly those derived from classified advertising, would almost certainly be dented by a universal viewdata system. Many other types of businesses would either be threatened or have to modify their practices. Travel agents and house agents (real estate brokers) might experience severe competition from a viewdata service, for example. Banking would almost certainly be affected, and professional services could eventually feel the impact.

Even elections and political events could be modified if, as seems possible, viewdata receivers should become as common as television sets now are. It is possible, for example, that after a political debate on television, viewers could be asked to record their opinions by answering a series of questions on the viewdata system. The system could certainly be used for rapid opinion polling and conceivably even for voting in elections.

These possibilities are, however, some way in the future. The more urgent questions facing the public relate to how the system should be allowed to develop in its early stages. Should information providers be allowed free access, as in the Prestel system? Or should codes and regulations be drawn up to define the potentially powerful uses of this medium? Should advertisers be allowed unrestricted competition with, say, public bodies and consumer organizations? How should the scale of charges be devised to be fair to the users, the information providers and the owners of the systems? How should commercial and advertising standards be enforced in the new medium? And perhaps most difficult of all, how can the immense possiblities for improving the knowledge and education of citizens be utilized and not perverted?

5

Videotext in the U. S.

by Efrem Sigel

The United States did not take the lead in developing videotext systems. That is clear from the preceding chapters describing the evolution of teletext and viewdata in Great Britain. The sheer size of the American market, however, and its pre-eminent position in the computer, telecommunications and information industries, made it inevitable that videotext would reach the U.S.

From 1976 to 1978, as videotext services got underway in Britain and France, U.S. companies and government officials watched first with indifference, then with quickened interest. Delegations of American visitors trooped off to the BBC in London to watch CEEFAX in operation; some of them moved on to France, Canada and Japan.

Meanwhile, the British and French realized that the investment made in equipment and computer programming could be recouped, at least in part, by licensing their systems for overseas use. The British Post Office mounted a vigorous campaign to sell such licenses, and a French government agency, SOFRATEV, quickly followed suit. Europeans began arriving in the U.S. to preach the gospel of CEEFAX, ANTIOPE, viewdata or other versions of videotext.

A 1978 study conducted jointly by Link Associates of New York and Butler Cox of London on the U.S. market for viewdata provided tangible evidence of the American interest: more than 40 U.S. and foreign firms signed up as sponsors, each paying around $8500 for the research. The roster included Xerox, McGraw-Hill, Dun & Bradstreet, John Wiley, RCA, Knight-Ridder, General Telephone & Electronics and many others.

By 1979 the stage was set for a number of companies to announce test versions of one or another videotext technology. The first half of that year saw at least seven such announcements, spanning every type of transmission system: over-the-air (broadcast) television, cable television, multipoint distribution service (a microwave common carrier service) and telephone lines.

The diversity of approaches from a bewildering array of organizations illustrates one striking difference between videotext in the U.S. and videotext in the European environment. Unlike the European countries, the U.S. has maintained both its broadcasting and telecommunications industries in the hands of private companies, albeit subject to close government regulation. Hence, no single decision of a government official could either set videotext in motion or halt it in its tracks. The government agency most directly concerned, the Federal Communications Commission, evinced notable interest, but its authority seemed limited to authorizing broadcast teletext services. Thus, it was apparent by mid-1979 that the pace of videotext evolution in the U.S. would be determined by the initiatives of private corporations, not by government actions.

In an attempt to bring some order to the variety of approaches in the U.S. the following sections are organized by type of delivery system: 1) broadcasting and related services, 2) cable television services and 3) telephone line delivery.

It should be emphasized that although the organization providing the *transmission facilities* (e.g., a broadcaster, a cable TV operator or a phone company) varies with the type of system, the organization providing the *information* can do so whatever the delivery system. Thus, a news wire service, a bank, a publisher of financial information and a company selling merchandise by mail order can each put its message on any type of system; several such companies are simultaneously exploring all three versions.

The other important qualification is that, as of mid-1979, no videotext service in the U.S. had any concrete experience. Apart from some technical tests of broadcast teletext, these services existed solely in the form of press releases and organization charts. Still, there was no reason to doubt that at least some of the services would in fact begin operating, and that 1980 and 1981 would be the critical time for proving their technical and perhaps economic feasibility.

BROADCASTING AND RELATED SERVICES

The size, wealth and reach of the U.S. broadcasting industry makes it a natural candidate for the adoption of teletext services. In 1978 there were 996 individually licensed TV stations in the U.S., 727 commercial and 269 nonprofit or educational. Ninety-eight percent of U.S. households are equipped to receive television transmissions, and 46% of the households—more than 33 million— have two or more sets. Seventy-eight percent of those homes have color sets, a significant fact in considering videotext services that are capable of full color display. Moreover, the television broadcasting industry remains one of the most profitable in the U.S. In 1977 its revenues reached $5.9 billion, and the pretax profits as recorded by the Federal Communications Commission were $1.4 billion.

The FCC authorizes stations to transmit through the license they are granted. It prescribes technical standards for transmissions, standards that also affect the TV receivers that are sold in the U.S. Any change in the nature of the TV signal transmitted, such as the insertion of teletext data in the vertical blanking interval, requires an explicit decision by the FCC setting technical specifications for such transmission.

Growth of Interest

Until the late 1970s, broadcasters paid little attention to the introduction of teletext services abroad. The only active effort to utilize this digital data technique in the vertical interval was started by the Public Broadcasting Service (PBS) in Washington, D.C. Its goal was to achieve a captioning system for the deaf or hearing-impaired viewer, who could dial up the subtitles on specially equipped TV receivers. Viewers with normal hearing would not be bothered, as their TV images would not contain the subtitles.

The digital data carrying the captions are inserted in line 21 of the TV picture and are invisible under normal circumstances. The FCC ratified the use of line 21 for this application in 1977; however the PBS system was just beginning its field test trials in 1979, and no major receiver manufacturers had begun making receivers equipped to receive and decode these signals. In fact, the PBS system is now being questioned in light of current interest in a full teletext system. Whereas the PBS approach uses line 21 for captioning only, other teletext systems are doing captioning by allocating only a small portion of their full capability to this function.

CBS, for one, has announced that it will not begin transmitting of closed captions because it feels such a service would waste spectrum space that could be used not only for captions but for other information. For example, in a service like the BBC's CEEFAX, just one page of the 100 pages per channel could be set aside for captions shown right under the normal program picture. Thus, viewers who wanted captions could have them, while there would still be 99 additional pages available for news, sports results and other information.

In 1978 efforts to implant a teletext or viewdata system in North America began in earnest. Both French and British representatives began to tour the continent giving lectures or hardware demonstrations. James Redmond, then director of engineering for the BBC, presented a luncheon lecture about CEEFAX at the March 1978 conference of the National Association of Broadcasters (NAB). Various members of the British Post Office participated in demonstrations of its viewdata system, now called Prestel.

There was similar activity on the part of SOFRATEV (Societe Francaise d'Etudes et de Realisations d'Equipements de Radio-diffusion et de Television), the French government agency responsible for promoting the ANTIOPE service. Jean Guillermin, president of SOFRATEV, demonstrated ANTIOPE at the National Cable Television Association (NCTA) conference in New Orleans in April 1978, at the International Communications Conference meeting in Toronto in June and at the Society of Motion Picture and Television Engineers (SMPTE) conference in New York in October.

Test Transmissions

On-air tests of teletext also began. With FCC permission, station KSL-TV in Salt Lake City began transmissions in June 1978 of a teletext service based on the BBC's CEEFAX. Since the 525-line television picture used in North America can display less data than the 625-line European systems, the teletext screen contained 20 lines of text, each containing 32 characters, compared with 24 lines of 40 characters in Britain and France.

To make the test possible, Texas Instruments agreed to provide modified TIFAX decoders (the decoder made by TI in Britain for

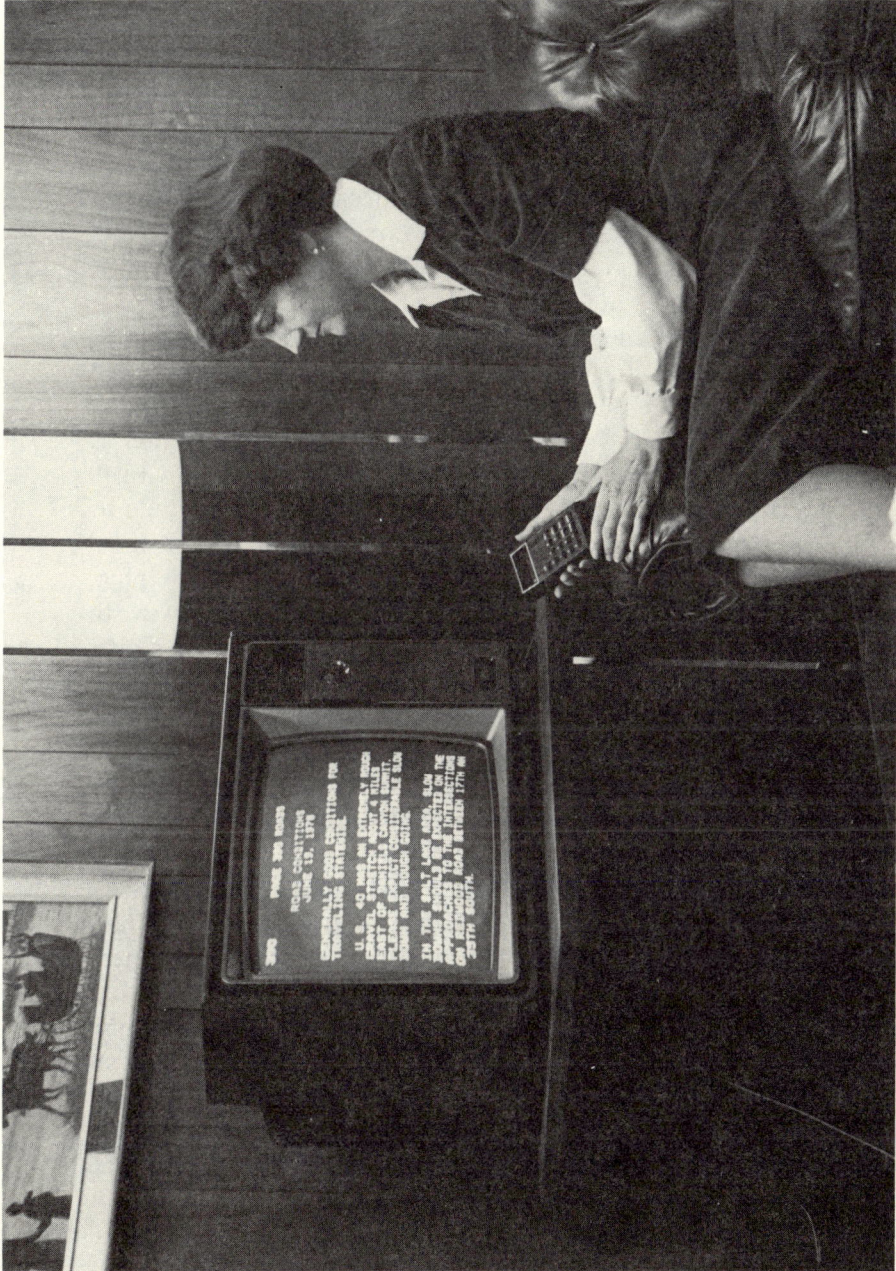

KSL-TV, Salt Lake City, began tests of a teletext service, based on the CEEFAX model, in June 1978. Courtesy KSL-TV.

the British teletext receivers) and Bonneville International Corp., which owns KSL, designed and built the computer-based encoder that generates the teletext signals.

According to Bill Loveless, director of engineering for Bonneville, the test used the following equipment: a computer encoder, a Tektronix insertion unit and a TIFAX decoder. The Tektronix unit inserted the digital information into lines 15 and 16 of the video signal. The decoder "contains a signal slicer, one page memory, video character generator, color matrixing circuits, timing and control logic," Loveless reported. These functions are provided by LSI (large-scale integrated) chips on a small circuit board.

Micro TV of Philadelphia, a subsidiary of Radio Broadcasting Co., came up with another version of teletext, which it called Info-Text. Micro TV developed its own combination of computer and related equipment for storing and manipulating the signal. It also used a different bit rate to provide a display of 20 lines with 40 characters per line. In certain cases, it said, up to 22 lines could be available. For its tests, Micro TV inserted teletext pages in the signal of its multipoint distribution service (MDS) station in Philadelphia.

MDS is a common carrier service that transmits by microwave to points within a specific geographic area. Transmission is on a higher frequency than is used for broadcast television, and special converters are needed to make the signal receivable on an ordinary TV set. Because MDS stations are common carriers, the FCC has no jurisdiction over the content of their signals. Hence, Micro TV could conduct its test—and would be free to introduce a commercial service—without government approval.

The KSL experiment, the Micro TV tests and growing interest in teletext generally prompted the FCC in 1978 to call for preliminary demonstrations that were closed to the public and the press. These in turn led to an open hearing at FCC headquarters in Washington.

On November 8, 1978, the full Commission, led by Chairman Charles Ferris, interrogated a panel of experts and saw representative hardware demonstrations of ANTIOPE, Info-Text and a tape of station KSL's version of CEEFAX. At the conclusion of this heavily attended open hearing, Chairman Ferris stated that the Commission would look favorably on any applications by broadcasters for experimental licenses to run teletext for on-air evaluation.

Robert O'Connor, speaking for the CBS network, indicated it

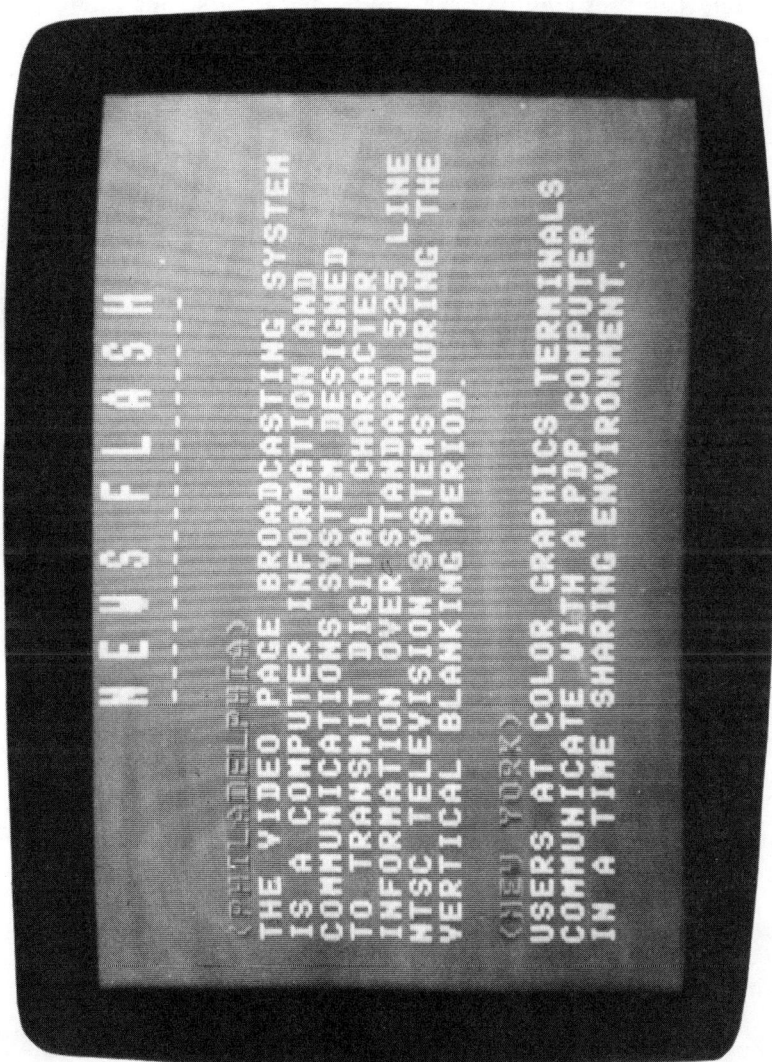

Micro TV's version of teletext, called Info-Text, transmits fixed graphic and alphanumeric material on the home TV screen without disturbing regular program transmission. Courtesy Micro TV Inc.

was interested in running such tests, and stated that it would petition the Commission for permission to do so.

In early 1979, the Electronic Industries Association (EIA) set up a committee on teletext to look into the standardization problems and to direct five task forces to run exhaustive tests on the teletext systems being proposed. At the first meeting of the Teletext Committee in Washington (February 13) all three TV networks and various television set manufacturers indicated their strong interest in establishing a common North American standard for teletext.

The CBS Television Network installed ANTIOPE and CEEFAX equipment at its owned and operated station KMOX-TV in St. Louis. The station began conducting parallel field measurements of the comparative characteristics of each system starting in March 1979. NBC also opted to conduct teletext trials using ANTIOPE at its Washington, D.C. station WRC. WGBH, the public TV station in Boston, expressed interest; KCET, Los Angeles, also did tests.

Interest of Manufacturers

As for the TV set manufacturers, they were represented on the various EIA task forces examining teletext. Zenith Radio Corp. supplied TV receivers for the station KSL test, and expressed its interest in manufacturing teletext receivers as a new product area. Nevertheless, as one Zenith technical manager explained, teletext receivers would only make sense if several concerns were met. These were: 1) to have standardization of displays from one delivery system to another, 2) to have a display that would carry as much information on the screen as possible (the 32 characters per line available in the KSL test was seen as insufficient) and 3) to have a product that could be offered at a reasonable price to the consumer.

A goal of adding no more than $100 to the retail price of the set was "not a bad number," this expert commented, noting that manufacturing quantities of 500,000 to 1 million a year would be required to make the effort worthwhile. Since total U.S. sales of color TV sets now are about 10 million units annually, this modest target represents a significant percentage of current sales.

Test of ANTIOPE

While broadcasters were edging toward technical tests of teletext, MDS operators were moving ahead on their own. Besides the Micro

TV tests (and Micro TV's activities in the cable TV industry, described later in this chapter) the largest MDS operator, Microband Corp. of America, also began exploring this technology. In May 1979 it reached agreement with SOFRATEV to conduct a joint test of the ANTIOPE system. The test would take place in the Washington, D.C. area using the Microband transmitter located in Bethesda, MD.

Mark Foster, president of Microband National System (a national network representing local MDS stations), explained the choice of ANTIOPE by saying, "We believe this is the best system— it has more flexibility and more opportunity." He specifically mentioned the ability of the ANTIOPE system to be used either for transmitting in the vertical interval of a regular TV picture or for sending data in all lines of the picture, so that the entire transmission would consist of teletext pages. In the latter application, users would be able to specify what items of information they wanted by using phone lines; the information would then be transmitted on demand, in portions of the bandwidth that had been set aside for those users.

Information to be transmitted by Microband during the six months of tests was to include stock market data, weather and crop information from the Agriculture Department's "Green Thumb" service, and information on the status of legislation in the House of Representatives.

Microband pointed out that the decision to test the ANTIOPE system did not rule out use of other technologies in the future. As a common carrier, an MDS station must be open to any user. Hence, a user who wanted to install CEEFAX-type equipment and transmit information in this format would be free to do so, though he would have to bear the cost of the equipment and receivers himself.

Economic Issues

While tests of the technical aspects of teletext were getting underway, broadcasters and MDS operators were saying little about the economic dimension. Broadcasters have no way to charge viewers for receiving teletext pages; one avenue they were obviously examining was to charge advertisers for sponsoring certain pages, or for including direct-response advertisements.

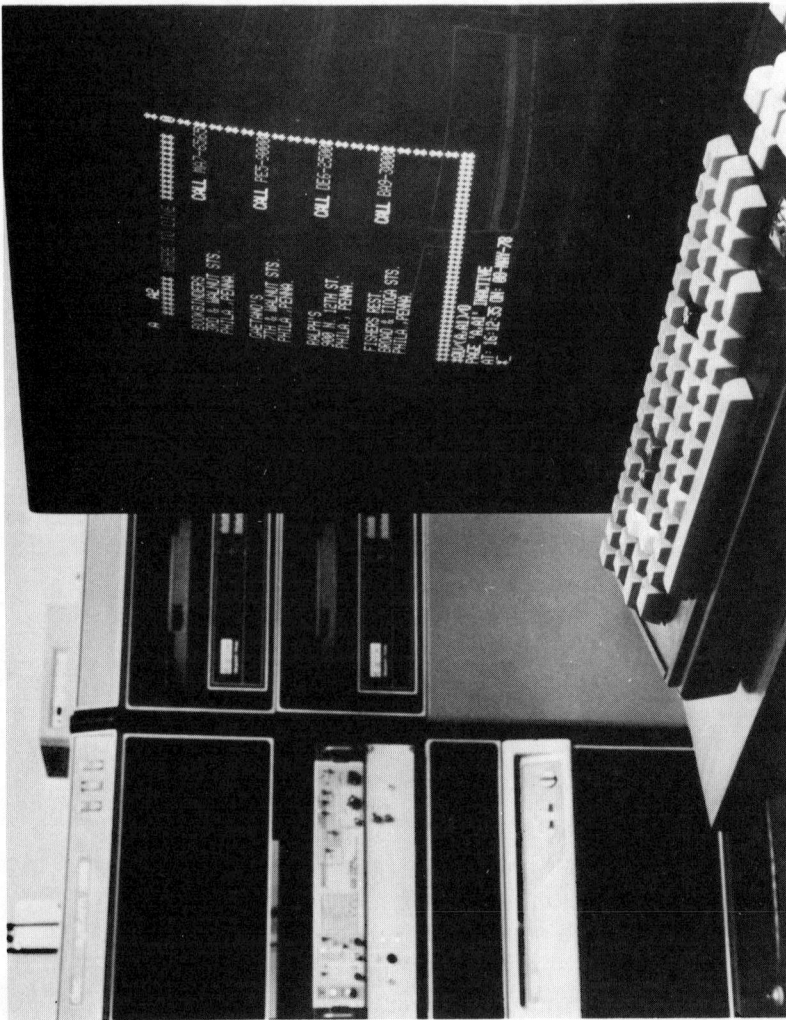

Micro TV's Info-Text encoder terminal (foreground) feeds graphic material into a computer that can store up to 3000 "pages" of material. Courtesy Micro TV Inc.

Another unresolved economic issue is the cost of producing teletext pages. This would be little problem for a national network like CBS or NBC, with its large news-gathering organization. Local stations, however, usually have very small news staffs responsible for putting on one or two broadcasts a day. Editing and transmitting a constant flow of news, information and listings would obviously call for increased staff.

MDS operators like Microband have a straightforward way of getting paid for videotext services: they rent their air time by the hour, and any user who wishes to make use of the equipment for sending and receiving videotext information could be charged for it. While the method may be straightforward, however, no one in 1979 had a clear idea of what the costs or charges would have to be.

With respect to the economics of videotext, the situation for cable television systems resembles that for MDS operators. Cable subscribers pay monthly fees for service received. Hence, the cost of a videotext service could be either added to the basic subscription fee or billed as incurred, if the cable system is equipped for two-way transmission. The following section describes the current status of cable television systems and their receptivity to videotext.

VIDEOTEXT AND CABLE TELEVISION

Only a tiny fraction of the 14 million cable television subscribers in the U.S. have the ability to send instructions or signals back to the operators of the system. Yet it is this technical possibility, called interactive or two-way cable, that gives the industry much of its allure. The coaxial cable that brings TV signals to a fifth of American households is what communications experts call a "broadband communications channel." This means it is capable of carrying many different transmissions at the same time. Cable systems built in the 1970s can deliver up to 36 different color TV channels. And since a color TV signal requires many times the bandwidth needed for data, audio or still pictures, this capacity translates into much greater potential for videotext or other information services.

Brief History

Cable has grown in fits and starts in the U.S. ever since the first community antenna system sprang up in the late 1940s to capture

Teletext experts showing equipment at FCC conference in November 1978. William Gross of Micro TV, Al Curll of Texas Instruments, Bill Loveless of KSL, François Renevier of SOFRATEV. Photo: Donna Foster-Roizen.

broadcast signals that were blocked by mountains. The first systems could only transmit a few channels and had no purpose other than to make possible reception of local broadcasts. Later, cable systems began using microwave hookups to import "distant signals" (those from stations not normally receivable in a local area). As recently as 1970, there were only 4.5 million cable subscribers, and the industry was an insignificant competitor to broadcasters as a way of delivering visual programs.

Three significant developments in the 1970s have transformed the cable industry: 1) the progressive lifting of almost all regulatory limits of what cable can offer, 2) the emergence of national pay TV networks that greatly increased the economic potential of the industry and 3) the development of a workable technology for interactive services on cable.

Emergence of Cable as Viable Industry

The lifting of regulatory limits, both through FCC orders and through court rulings, has freed cable operators to bid for and air first-run movies, to import distant signals and to offer or not offer free community channels as they wished. The emergence of national pay TV networks, principally Home Box Office (owned by Time Inc.) and Showtime (owned by Viacom and Teleprompter), has added enormous revenues for cable operators. By mid-1979 more than 3 million households were subscribing to special pay cable channels, forking over an average of $8 per month for the privilege. Beyond the welcome revenues to cable operators, the pay cable networks and the movie companies that supply most of the programs, the service has accustomed consumers to paying extra for additional services. At present that additional service is limited to entertainment. A key, still unanswered question is whether consumers will pay extra for information services, too.

The third development, the production of workable hardware for two-way signals, scarcely rates as a momentous technological breakthrough, since the techniques have been known for years. The breakthrough that is required is an economic and marketing one. It consists of finding the right mix of services that will attract enough customers to pay for the cost of investing in added transmission equipment.

According to Robert G. Vallerand, engineering vice president for the western region for American Television & Communications, there are three basic designs for a cable system that can carry two-way signals.

One-way and Two-way Technology

The simplest uses a one-way cable in the 50 to 300 Mhz bandwidth to carry the basic video signal to subscribers. A return line in the 5 to 30 Mhz bandwidth can carry a signal back to the "headend," or point where the signal originates. The capital cost of this type of system is about $6000 per mile, compared to $4000 per mile for a basic cable system that cannot handle two-way traffic.

A more versatile design incorporates two trunk lines for carrying forward signals to subscribers, one of which can also handle reverse signals from subscribers back to the headend. Such a system would cost $9000 per mile.

The most sophisticated design of all provides two complete systems, A and B, each with the ability to carry two-way signals. The cost of this version would be $12,000 per mile, double the expense of the simplest two-way design.

Whatever the design of the transmission system, two-way cable requires a special terminal in the subscriber's home that will send signals to a central point. In the simplest of systems, this terminal can consist largely of sensors that detect any change in a normal condition: e.g., excessive heat indicating a fire, or the interruption of a circuit that would signal a burglary in progress. More sophisticated two-way terminals incorporate a keypad that enables the viewer to pick one of several options. For example, he can order a movie or other program (for which he pays a special fee), give a yes or no opinion on a local question, order a tennis shirt or a clock radio from a merchandise catalog displayed on the screen.

Such a terminal sends the signals over the return line in the cable system to a computer located at headquarters. Each terminal has a unique identifying number so that, in the case of the movie or the item of merchandise, the person ordering it can be sent a bill.

Each of these services was actually available in a cable system in 1979, though the examples were still being counted by ones and twos. The passive use of a two-way system for fire and burglary alarms was demonstrated in Rockford, IL, where the equipment was installed as part of a project funded by the National Science Foundation. Another cable system, Woodlands CATV in Woodlands, TX, offered subscribers a terminal made by Tocom that can signal fire, burglary, medical emergency and other problems. The terminal cost about $300.

A more active use of two-way cable was for communication between several offices of the same organization. Bankers Trust used the Teleprompter system in Manhattan for sending computer data back and forth between the main office and the data processing center. The American Television & Communications franchise in Durham, NC, served a similar function for IBM facilities in the area. In these applications, special interconnection equipment is installed by the organization that needs the two-way link. Other users, however, are prevented from making use of this feature of the system.

Warner's Qube

The most ambitious example by far of two-way cable is the Qube service launched by Warner Cable in Columbus, OH, in 1977. In this system, subscribers can play game shows or quizzes, register their approval or disapproval of participants in a local talent show, "vote" on community issues (votes are advisory only) or order movies and other entertainment programs. Market research and direct selling have been another early use of Qube.

This last feature pinpoints one of the most troubling aspects of two-way cable. Because each subscriber has his own ID number and a computer sweeps each home every six seconds, the service can accumulate vast amounts of information on what its customers are watching, how they're voting and what they're buying. Warner officials disclaim any intention to maintain records on individual viewing or buying habits. But they acknowledge that concern about privacy is genuine, and that the potential for abuses cannot be taken lightly.

Videotext via Cable

Conventional one-way cable offered a variety of information services in 1979, and there were glimmerings of broader services. One channel was usually allocated to giving the time and weather, another to providing news headlines and summaries, courtesy of United Press International or the Associated Press. The latter is a standard type of newswire in which the order of information is set

Subscribers to Warner Cable's QUBE registered their opinions of a speech by President Carter, answering questions put to them on live television by NBC correspondents. Subscribers' reactions were broadcast across the country. Courtesy Warner Cable.

by the news agency, and the viewer must wait for the information he is interested in to come on the screen. Moreover, an entire video channel must be devoted to weather or news headlines or sports results.

A logical development would be to marry the teletext technology for carrying digital information in a broadcast signal to cable's wealth of available channels. Two possibilities are available. One is to use the vertical blanking interval of an existing signal to send 100 or so pages of current information. Another is to set aside an entire channel for videotext, with the potential for 25,000 pages being continuously recycled and available for retrieval.

One of the first companies to explore teletext on cable was Reuters. In 1979 it was already operating a simple news service called News-Views that went out by phone lines to subscribing cable systems. They used a character generator to create text pages sent out over either one or two cable channels. At the National Cable Television Association meeting in May 1979, Reuters announced a version called News-Views by Satellite. This service involved transmitting the pages in the vertical interval of an existing broadcast signal being carried on the RCA domestic satellite. Subscribing cable systems would decode the teletext signal by means of a special terminal, then shunt the signal to a channel set aside for this related News-Views service.

In both News-Views services, the Reuters information consisted of about 30 pages of general news and financial information. Pages were sent in rotation, and there was no way for a subscriber to ask for and freeze a specific page.

Another Reuters service, The Reuter Monitor, uses an entire video channel to transmit much more detailed financial information to business subscribers. In this service subscribers have a decoder to call up just those pages they are interested in. The service was available in 1978 in one cable system in the New York metropolitan area: Manhattan Cable.

To make the service available nationally, Reuters rented a transponder (a communications circuit equivalent to one color video channel) on the RCA satellite beginning in 1979. Its intention was to send the signal out by satellite from New York, then pipe it into subscribers' homes or offices in other cities either by a cable system or by a local MDS station.

The Reuter Monitor illustrates Reuters' commitment to computerized information retrieval, but it is entirely separate from the News-Views by Satellite service announced in 1979. The transmission channel and hardware making possible News-Views by Satellite came from SSS-CableText. This was a joint venture of Satellite Syndicated Systems of Tulsa, OK, and Micro TV of Philadelphia. SSS, a common carrier, owns earth stations and rents transponder channels on the RCA satellite. It carries the signal of WTCG (channel 17), Atlanta, to more than 800 cable TV systems around the country.

Micro TV used its experiments with teletext, and the equipment it had developed, as the basis for its entry into the cable television market. It was to be responsible for supplying the decoders that cable operators would need to display the Reuters service. Since Reuters would occupy only 30 of the 100 or more pages available in the SSS-CableText service, the organizers expected to sign up other news organizations or publishers that would make use of the transmission channel to serve cable operators.

Each information supplier would pay for storing its pages in the SSS-CableText system, while each cable operator would pay to rent the decoders. Information suppliers could also charge the cable operators whatever they wanted for the information itself; Reuters expected that this fee would be less than $300 per month.

Exhibits at the NCTA convention in Las Vegas gave evidence of the sudden interest in videotext by the cable industry. Micro TV and Satellite Syndicated Systems showed their SSS-CableText service. SOFRATEV again exhibited the ANTIOPE system. And a leading manufacturer of cable television equipment, Oak Industries of Crystal Lake, IL, announced a forthcoming service that it called "total control videotext."

Oak had previously developed what are called "addressable" converters for cable television and over-the-air subscription TV. "Addressable" means that the converter can be turned on or off for reception of a particular program, by sending an electronic signal from the headend of the cable system to the unit located in a customer's home. This feature eliminates the need for a serviceman to install a converter, or to remove it if the subscriber moves or fails to pay his bills.

Total control videotext was to combine this addressable feature with a broad range of videotext information services. Oak's intention was to sell cable operators a minicomputer and related equipment to store and transmit textual information from the cable system headend to subscribers. Signals could either be inserted in the vertical interval of an existing program, or an entire channel could be set aside for videotext, with much greater capacity in pages.

Oak could not say what information would be transmitted on this system, nor whose equipment would be used. Minicomputers made by IBM, Digital Equipment Corp. and Data General were all under consideration. Decoders, to be placed in individual subscribers' homes, would be made by Oak using microprocessor "chips" supplied by a semiconductor company.

Oak spokesmen were unable to specify what display format would be used for this service, when introduced, though one possibility was to have 32 characters per row by 20 rows. The earliest that Oak would have a system for installation by cable operators would be in 1980.

Conclusion

The arrival of videotext technology in the late 1970s coincided with cable television's new-found prosperity, as a result of increased subscriber penetration and the upsurge in pay television subscribers. Moreover, where cable TV systems were built with the technical ability to carry two-way signals, or where such an ability could be added, the systems offered a ready-made network for the most sophisticated variety of videotext: a viewdata service in which customers could send signals to a computer to order just the information they wanted, when they wanted it.

Few cable system managers had any knowledge of videotext as of mid-1979, however, nor were they experienced in the organization and marketing of information services, as opposed to TV entertainment. Thus it will be up to outside suppliers like Reuters and Oak to develop the specific services, rent channels on available cable systems and undertake the lengthy, costly process of introducing these services to prospective customers.

One indication that videotext services on cable may not be universally popular comes from a survey conducted for the National Cable Television Association in June 1979 by Peter Hart

Research Associates.* Of eight types of programs or services that might be available on cable, a videotext type of service ("a channel which would provide teletype service on stocks, consumer information and weather") ranked the lowest, although 30% of respondents deemed it important. However, 75% ranked first-run movies and concerts as important, 54% ranked sports programming important, 52% children's programming, etc.

It can be argued that since few people have been exposed to the variety of information possible with videotext, they have little understanding of its true appeal. More likely, however, is that for most people, television is firmly established as an entertainment medium, not an informational one.

The following section considers viewdata services that can be offered over ordinary telephone lines.

VIDEOTEXT BY TELEPHONE LINE

While broadcasters and equipment manufacturers circled warily around teletext, and cable operators got an initiation into videotext technology, a number of companies were quietly laying plans for viewdata services involving computer storage of information and its transmission over telephone lines.

By mid-1979, at least three separate ventures had been announced. An analysis of their sponsors shows the mix of organizations that are interested in this technology: news organizations, a computing company and a giant telecommunications common carrier.

Knight-Ridder Newspapers, Inc. was the first to make an announcement. At its annual shareholders' meeting on April 17, 1979, president Alvah Chapman revealed establishment of a subsidiary called Viewdata Corp. of America to develop an electronic home information system, to be called Viewtron.

The Viewtron service was to involve transmission of news, sports results, calendars of local information, lists of adult education courses, movie, restaurant and theater schedules, and other useful facts. The information would be stored in the Knight-Ridder computer center in Miami, FL, and transmitted over existing phone lines to homes equipped with a specially adapted TV set. The initial test was to begin in 1980, with 160 families in Coral Gables, FL.

---

*"A Survey of Attitudes Toward Cable Television," Washington, D.C.: Peter D. Hart Research Associates, Inc., 1529 O Street, N.W., Washington, D.C. 20005.

In researching the new service, Knight-Ridder staff members visited systems in England, France, Germany and Japan. Significantly, Chapman called the service "an adaptation of the British viewdata concept to the U.S. market." To gain experience in interactive home information systems, Knight-Ridder is participating in the British Prestel service as an information provider. The information it is supplying consists of facts of interest to Miami tourists—information about the city's weather, geography, transportation, medical services, fishing and outdoor sports, etc. It is no accident that much of the same type of information would be useful in a data bank aimed at serving local subscribers, too.

Chapman named three top officers of Viewdata Corp. of America: Albert J. Gillen, president of Knight-Ridder Broadcasting, is to serve as president. Hal J. Jurgensmeyer, a senior vice president for planning, was appointed senior vice president of the new subsidiary. Norm Morrison, vice president for research and production of KR, was named vice president of Viewdata Corp. of America and manager of the pilot project. Knight-Ridder said that the project would cost approximately $1.3 million over two years, and that the new company would employ about 30 people. AT&T agreed to cooperate with the test. Among the publishers and advertisers that signed up to put information into the computer were the Associated Press, Consumers Union, Congressional Quarterly, the Miami *Herald,* Eastern Airlines and Shell Oil.

The significance of the KR announcement was obvious. Knight-Ridder is the largest newspaper chain in the U.S. in terms of number of subscribers; its 36 daily papers reach more than 3.7 million subscribers a day. Among the major dailies published by the company are the Miami *Herald,* the Philadelphia *Inquirer* and the Detroit *Free Press.* Except for its broadcasting division, Knight-Ridder has remained strictly a conventional newspaper publisher, so its decision to experiment with viewdata signifies the powerful attraction this technology has for newspaper publishers—either that, or the threat it represents to their pre-eminent local position in disseminating information.

Telecomputing Corp. of America in McLean, VA, a company owned 10% by Dialcom, 80% by Digital Broadcasting Co. and 10% by other investors, was the second organization to announce its entry into electronic home information. TCA arranged a joint venture called News Share with a major news organization, United Press International. UPI news, sports and features would be stored in TCA's data center in Silver Spring, MD. The service would be available to subscribers who own a home computer or who rent special terminals: they could dial up this data base and select specific items of information to retrieve. TCA will bill subscribers by connect time—$15 an hour during business hours, $2.75 an hour after 6 p.m. The two companies will try to sign up local papers around the country to store local news and classifieds in the data base, so that subscribers can retrieve items about their own community. Because existing off-the-shelf computer terminals can be used in this system, no investment is required to design and manufacture special terminals. On the other hand, that limits the service basically to black-and-white display, without graphics, whereas the service that Knight-Ridder announced—based on the British model—will provide full color and simple graphics.

In June 1979, two months after the UPI announcement, TCA reached agreement with a second information supplier. The New York Times Co. is providing a modified version of its Information Bank through the Telecomputing system. Other suppliers of information in the data base include Prentice-Hall, which is providing tax information, and Media General, publisher of the Media General Financial Weekly, which had agreed to supply financial information on 5000 publicly owned companies. By fall 1979 there were 500 million characters of information in the TCA computer. TCA called its service "The Source."

In September 1979, TCA chairman William von Meister reported there were a couple of thousand subscribers to the service, of whom 20% were computer hobbyists and 80% were business organizations. TCA was marketing it through advertisements in personal computing magazines, as well as through sales calls on businesses. It also planned to market through local franchises; the first three such franchises were located in Washington, D.C., Atlanta and Colorado Springs, CO. According to von Meister, the early subscribers were dialing up the service an average of eight to 10 hours per month.

General Telephone & Electronics was the third major organization to enter the viewdata field, announcing its service in June 1979. GTE obtained a license for the British viewdata service, Prestel, through Insac Viewdata, Inc., the U.S. representative of the British Post Office.

Unlike its two rivals, GTE gave only the barest details of what it intended. GTE's spokesman was Lee L. Davenport, vice president and chief scientist. He said the company's initial effort would be "market testing activities," adding that "a number of leading companies have already expressed interest in participating in this service, including organizations prominent in the publishing, financial, news and entertainment fields."

As a common carrier and a manufacturer of television and electronic equipment, GTE has a different orientation from that of Knight-Ridder or UPI. Rather than establishing itself as the owner and supplier of information, GTE would presumably wish to involve a large number of information suppliers, in order to give any service the widest possible appeal. Such appeal would stimulate demand for both the communications circuits that the company provides and the equipment needed to receive viewdata transmissions.

Insac Viewdata, the firm from which GTE acquired its license, had been holding discussions with U.S. companies for more than a year before the contract was signed. Insac has modified the computer software used in Britain for American applications. One application that Insac has explored is using viewdata for communication among closed user groups: by entering a special code, employees of a company could send and receive private messages, members of a professional society could obtain membership notices, etc. Another application is transaction services, like banking and shopping. In an interview in *VU Marketplace,* a trade newspaper, Insac president John Bately specifically mentioned the appeal of viewdata in permitting a customer to consider and purchase an item from a TV display.

Other U.S. companies are waiting in the wings, their entry into home retrieval services seemingly only a matter of time. The New York Times Co. is one: its Information Bank is a vast computer data bank providing news information to subscribers in government, industry and libraries. The Times was one of the first U.S. firms to put some of its information on Prestel, the British system, as a way of testing the service. The trial with Telecomputing Corp. of America is another experiment, but the Times is free to market its own data base itself—or through other middlemen.

Dow Jones is another potential supplier. Its computerized

information retrieval service, News/Recall, is available through terminals in the business and brokerage house community. Already certain home computers can tap the Dow Jones data base, and DJ executives have carefully studied the potential of teletext and viewdata services.

It is far more likely, however, that business and professional audiences will be the first customers for videotext services. And there are scores of companies that have data bases of business information that can be transmitted in this way: A.C. Nielsen, which supplies grocery product information and TV ratings; Dun & Bradstreet, with its credit ratings and market statistics; IMS International, a supplier of information on medical/pharmaceutical industry trends; McGraw-Hill, with its Standard & Poor's financial services and its F.W. Dodge construction information; Data Resources, with its computer models that forecast changes in the U.S. economy; and many other publishers, information companies and associations.

In addition to these publishers, large computer and communications companies are intensively investigating computer-based information services. AT&T, IBM and Xerox are the most notable examples. AT&T has proposed a system it calls ACS, Advanced Communication Service, that would provide new transmission channels for computer data.* IBM, through its investment in Satellite Business Systems, is building communications channels that can carry information directly from one office to another, by means of satellite, without ever using phone company facilities. It is also researching a host of information technologies for the home, and in 1979 announced establishment of a joint venture with MCA to manufacture video discs and disc players for the home. Xerox, for its part, has made office automation a specific target and has also filed application with the Federal Communications Commission to build its own common carrier facilities under the name XTEN, Xerox Telecommunications Network.

---

*In August 1979, AT&T began a limited test of electronic retrieval of directory listings, by installing computer terminals in 75 to 100 homes and offices in the Albany, NY, area.

SUMMARY

After a slow start, videotext technology in all its forms is attracting lively interest in the U.S. As of mid-1979, a number of broadcasters were conducting tests of experimental teletext transmission, while equipment manufacturers studied problems of specifications and standards. The Federal Communications Commission must ultimately decide on nationwide standards.

Tests were also underway using multipoint distribution service, a common carrier service relying on microwave transmission. And several companies had announced videotext services aimed at cable television operators, which would use spare cable channels, or would insert a teletext signal into an existing program.

The most ambitious projects were underway in viewdata services that will use telephone lines for direct communication with a computer. Both Knight-Ridder Newspapers and General Telephone & Electronics announced services in the spring of 1979, with a number of other major companies likely to follow.

6

Videotext in Other Countries

by Joseph Roizen

While Britain took an early lead in teletext and viewdata, and while the U.S. was just beginning to investigate these technologies, other industrialized countries around the world were conducting their own research. France and Canada devoted considerable resources to videotext through national research centers or tele-communications agencies. Work was also underway in Japan and in several other European countries.

All systems share certain common features or are extensions of the original U.K. and French versions. Signals carrying the data are in digital form; transmission is via television or telephone channels; display devices are color TV receivers equipped with teletext decoders and, in the case of viewdata, with phone line couplers known as modems.

Standardization is being sought by various international organizations that deal with these problems, but as yet, no universal standard of signal format has emerged.

The current status of teletext and viewdata in various countries is as follows:

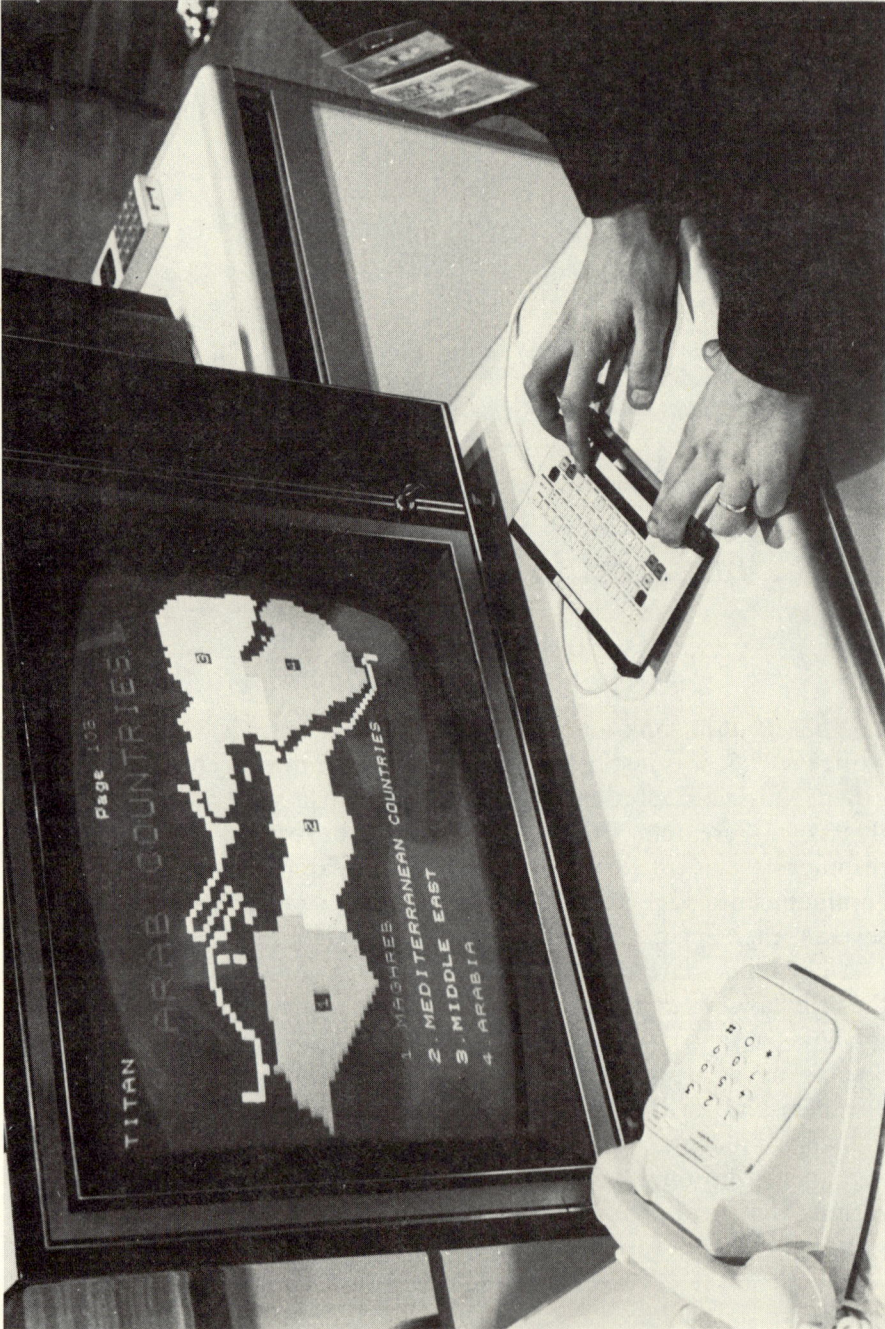

TITAN interactive system as exhibited in Dallas, 1978. A phone line to Rennes, France, served as the communication link between Dallas and the remote data base. Photo: Donna Foster-Roizen.

FRANCE

The French took a somewhat different approach to the development of teletext and viewdata systems. This was the result of having a research center in Rennes, the CCETT (Centre Commun d'Etudes de Télévision et Télécommunications), which is jointly supported by TDF, the national TV network, and the PTT, which operates telephone services.

The basic concept is a hybrid system that can cater to both teletext and viewdata needs, while sharing much of the same hardware and software. The French teletext system is called ANTIOPE (Acquisition Numérique et Télévisualisation d'Images Organisées en Pages d'Ecriture), an acronym that is also the name of a Greek goddess. The viewdata system is called TITAN and is fully compatible with ANTIOPE. Both systems use a digital signal packaging and insertion system known as DIDON. The latest nomenclature for the complete French system is ANTIOPE VIDEOTEX, and it has been operating in France since May 1977 when the Paris Bourse (stock exchange) went on the air over the national TV network with stock prices and other financial data.

ANTIOPE is now in use as a teletext service in Paris and Lyon. It has added weather pages to its magazine, and there are current plans to expand the ANTIOPE teletext services to include news, sports, classified ads and other topics of interest to viewers.

It is interesting to note that the stock quotations (up to 80 pages) that are broadcast by ANTIOPE are transmitted automatically without the need for manual keyboard entry of the constantly fluctuating data. An IBM 370 computer, which maintains the Paris Bourse statistics on an up-to-the-minute basis, is directly linked with the DIDON multiplexer, which in turn feeds the data stream to the TV transmitter.

Stockbrokers, financial houses and other interested parties equipped with color TV receivers that have ANTIOPE decoders can dial up stock market information as timely as that obtainable on the floor of the Bourse.

As for interactive viewdata services, ANTIOPE VIDEOTEX is also being developed. To fully test the potential of an interactive system, French authorities plan to install terminals in 3000 households in the town of Velizy in October 1980.

ANTIOPE display showing range of character sizes and ability to handle the Cyrillic alphabet—lower right.
Photo: Donna Foster-Roizen.

Since the ANTIOPE system offers some unique features as a combined teletext and viewdata service, it is being promoted vigorously in North America, Continental Europe, Asia and Australia, and has been used or proposed for such special international events as the World Cup Matches in Argentina (1978) and the Moscow Olympic Games in 1980. In the latter case, reporters would be able to look up background statistics via TV sets equipped with ANTIOPE decoders and linked to a computer.

The organization in France charged with the promotion of ANTIOPE VIDEOTEX is a subsidiary of TDF known as SOFRATEV.

The ANTIOPE advantages claimed by the French developers include:

- Greater flexibility due to common technology in teletext and viewdata applications.
  Greater precision in addressing information to specific groups of viewers, as well as full TV channel utilization for high volume, high speed data transfer.
- Easy adaptability to the various TV standards around the world. ANTIOPE can be simply adapted to SECAM, PAL or NTSC systems, or to the scanning rates they use.
- Common hardware should lead to greater volume, thereby achieving lowest possible prices for ANTIOPE display terminals.

CANADA

Next to the British and the French, the Canadians have made the largest governmental commitment to videotext systems.

Since the mid-1970s, the Communications Research Centre, a research arm of the federal Department of Communications, has been developing a system for interactive use over phone lines or cable, or for one-way broadcasting. The system, known as Telidon, is distinguished by its ability to reproduce complicated graphics with a high degree of fidelity.

The developers of Telidon have stressed that their system permits the display of information on the user's screen to be independent of the way the data are stored in the computer. They refer to the display feature as "alpha-geometric," as contrasted with the "alpha-

ADVANTAGES OF CANADIAN VIDEOTEX OVER PRESTEL/ANTIOPE

- VASTLY IMPROVED IMAGE QUALITY
  Resolution is whatever terminal permits.

- INFORMATION STORAGE INDEPENDENT OF TERMINAL DESIGN
  Future generations will still be able to access today's data.

- EFFICIENT USE OF COMMUNICATIONS MEDIA
  Actually improves with increase of display resolution.

- SIMPLE INFORMATION GENERATION PROCEDURES

- ADAPTABLE TO PERSON-TO-PERSON IMAGE COMMUNICATIONS

Canadian developers of Telidon feel their system is superior to the British Prestel or the French ANTIOPE. This screen display shows the advantages claimed for Telidon. Courtesy Canada Department of Communications.

mosaic" display of other videotext systems, such as Prestel. The Telidon terminals already demonstrated can provide 320 x 280 picture elements, compared with approximately 60 x 80 elements in other videotext displays. Higher definition Telidon terminals are under development in the laboratory.

Because the microprocessor circuitry in the terminal can be used to create geometric shapes on command, the computer where information is stored need only send, for example, the beginning and end points of a line; the terminal itself "draws" the line on the basis of this information. (Some additional memory is needed in the terminal compared to other videotext systems, but Telidon developers say this will not be costly.) Graphics are of much higher quality than in other systems. In addition, photographs can be sent by having the TV camera scan the image, which is then converted into digital form for storage and transmission.

The following Canadian organizations were planning to conduct tests of Telidon in late 1979 or in 1980:

- Bell Canada had 25 terminals operating in a test of a service modeled after the U.K.'s Prestel in early 1979, but had added terminals that would handle Telidon by July 1979. Three newspaper companies participating in the test—Torstar, Southam, Inc. and F.P. Publications—announced a joint venture to put their information in the Telidon format. The three said they would also convert information from other publishers and suppliers to the Telidon format.

- The Ontario Educational and Communications Authority planned tests of Telidon both over phone lines and in the vertical interval of a broadcasting signal. The latter test was to take place over TV Ontario, an educational TV network for the province.

- Alberta Government Telephone, Manitoba Telephone (both provincially owned phone companies) and British Columbia Telephone (a subsidiary of General Telephone & Electronics) also planned trials of the Telidon system.

- One cable television company, Telecable Videotron of St. Hubert, Quebec, planned trials of Telidon on its cable system.

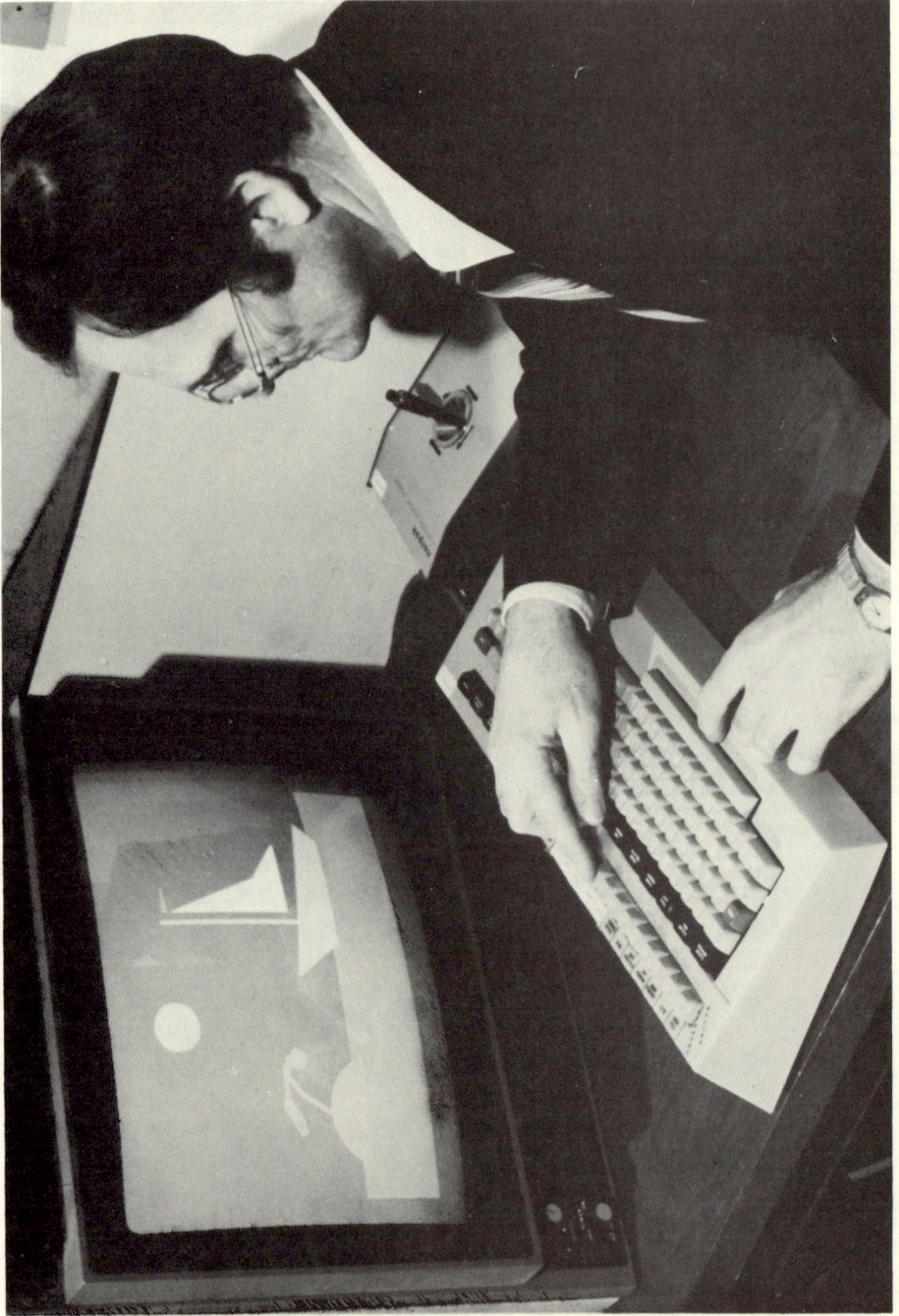

Telidon display showing the system's ability to display graphics. Courtesy Canada Department of Communications.

Several Canadian manufacturers were involved in making terminals and other equipment for Telidon. The principal supplier is Norpak, Ltd., of Pakenham, Ontario. S.E.D., of Saskatoon, Saskatchewan, had also manufactured terminals. Northern Telecom is another potential supplier, while Mitel, an electronics company in Ottawa, had a development contract with the Department of Communications to build a microprocessor "chip" for a Telidon terminal using very large-scale integrated (VLSI) logic.

The investment in Telidon by the Canadian government had amounted to $3 million by early 1979. Another $3.2 million was being spent by the government in 1979, with a further $6 million being sought over the next two years.

Canadian officials recognized that the selection of videotext systems in the U.S. would largely determine North American standards, given the size of the American market in both broadcasting and telephone distribution. This is particularly true in broadcasting. Since Canada and the U.S. have the same broadcasting standard (NTSC), and since many Canadian homes receive American telecasts either over the air or by cable, standards ultimately adopted by the FCC in the U.S. will probably determine the Canadian system as well. According to Douglas Parkhill, assistant deputy minister, research, of the Department of Communications, the developers of Telidon "have been talking with all of the major companies and carriers" in the U.S. regarding trials of Telidon technology.

Of all the Canadian organizations testing videotext systems, Bell Canada has the most ambitious plans. It gave the name Vista to its service, whose initial trial got underway in early 1979. Vista was originally modeled after the British Prestel system, but with a computer system and terminals assembled by Bell's own engineers from commercially available equipment. Subsequently, Bell began installing Telidon terminals for information providers, while planning for more widespread public tests of an interactive service. These tests were scheduled to take place in 1981 in about 1000 homes in the Toronto area. The Department of Communications agreed to provide 675 Telidon terminals for the test, with Bell Canada purchasing the other 325. Of the $10 million cost of the test, the DOC would underwrite $2.5 million and Bell $7.5 million.

Bell sees its ultimate role in videotext as one of benefitting from increased phone usage, and also as providing hardware through its manufacturing subsidiary, Northern Telecom.

Though there is occasional confusion as to whether Telidon and Vista are competing services, officials at the Department of Communications and at Bell Canada explain the distinction carefully. Telidon is a technology, not a commercial service; the federal government is developing the technology for use by Canada's own communications industry, as well as for export. Vista is simply a trade name for a service that can use any one of the videotext technologies, including but not limited to Telidon.

JAPAN

Both teletext and viewdata-type systems have been under active development in Japan by government agencies and private industry.

The Japanese TV network NHK (Nippon Hoso Kyokai) maintains a large research center on the outskirts of Tokyo known as the NHK Technical Research Laboratories. In February 1976, NHK published a proposal for a Text Television System. This paper also outlined one of the fundamental problems of a teletext or viewdata system in Japan. The Japanese written language needs more than 2000 pictographs or symbols to convey information of any great range, in comparison to the 26 letters of the Roman alphabet. This imposes a very severe limitation on how much real information a Japanese television text system could convey to viewers.

Nevertheless, NHK went on the air late in 1978 with an experimental teletext system which could provide such diverse services as captioning for the deaf, weather forecasts, shopping and TV program guides, etc.

There are in fact a number of separate projects going on in Japan that relate to teletext or viewdata. The NHK system already referred to is called the CIBS (Character Information Broadcasting Station) teletext service. This system is based on a matrix of dots with 332 dots per line and using 200 lines. A complete display of Japanese characters or a graphic is composed of almost 70,000 bits. The system is quite different from the U.K. or French systems, and requires a substantial memory in the receiver. It does use the vertical blanking interval for transmitting the data.

As for viewdata, the Japanese postal and telecommunications agency (PTT) has developed a system called the Character and

Japanese teletext. This display shows why the technical problem is much greater for a non-alphabetic language. It takes 2000 ideographs to convey messages that the English language does with a 26-character alphabet. Photo: Donna Foster-Roizen.

Pattern Telephone Access Information Network System (CAP-TAINS), which resembles the U.K. Prestel system. In conjunction with a variety of companies that have data bases, the PTT has set up a computer center in Tokyo and is conducting experiments that involve about 1000 households. The trial services provided to these users include educational material, community affairs information and general living data like health care information, cooking recipes, shopping specials and nearest facilities for various family needs.

Asahi Broadcasting in Osaka, one of Japan's major independent TV networks, has also developed a system it calls Telescan.

Telescan lets the viewer choose a variety of messages that appear as a crawling text superimposed on a regular TV image. The data bits for Telescan are also in the vertical interval and not visible under normal viewing conditions.

Several other two-way TV communication systems are in experimental use in Japan. The Tama New Town's Co-axial Cable Information Service was implemented in 1976 and has been expanded to give this localized group of viewers access to community antenna signals, pay TV services, facsimile newspapers, still images on request and other information. VRS is another interactive system set up by Nippon Telegraph and Telephone. Its major feature is an information center equipped with microfiche, 16 mm telecine and other image or sound sources that can be accessed by viewers. Hi-OVis is yet another experiment, this one set up in the Nara Precinct and using about 350 kilometers of optical fibre for two-way communications.

For all this effort, little of the Japanese developments in this field have had much effect yet in the outside world. As teletext and viewdata develop, no doubt products designed for the domestic Japanese systems will begin to find their way into more general use.

WEST GERMANY

The Berlin Radio Show, one of the world's largest electronic expositions, has been the site of major displays of both the British and French teletext and viewdata systems. By 1977 the major networks in Germany had participated in teletext demonstrations and the German Federal Post Office (Bundespost) had become involved in viewdata. On the teletext side, the German Institut fur

NHK's teletext system could be used to transmit illustrated children's stories. In this example, the images are created by a computer-based graphics generator. Photo: Donna Foster-Roizen.

Rundfunktechnik in Munich was charged with running field trials on a CEEFAX-based system.

The Bundespost planned a market trial of its viewdata service in 1980 in the Düsseldorf/Neuss area involving some 2000 subscribers. Another test was planned for West Berlin.

HOLLAND

The Dutch national TV network NOS has been experimenting with the CEEFAX system to determine its utility in Holland. Receivers made by Barco which contain teletext decoders have been used to display the teletext messages. These receivers are of a very advanced type using an infrared pulse transmission system for remote control of the TV set. The range of teletext options includes buttons for hidden information (puzzle solutions), timing of teletext switchover, or monochrome superimposition of teletext on a color picture. Viewdata in Holland is also being investigated by the Netherlands postal and telecommunications agency, which acquired a Prestel system for test and evaluation. Holland has not yet announced any firm decision on when public or private services will commence.

SCANDINAVIA

Sweden, Denmark and Finland are all involved in various tests with teletext or viewdata systems based mainly on the U.K. models. In Sweden the main effort has been to provide captions for the deaf as well as subtitles in southern European languages for immigrants not fluent in Swedish. The teletext system in Sweden is called Extratext. Sveriges Radio conducted a trial through its Audience Research Group, which allocated 160 teletext receivers to people with impaired hearing. According to surveys by the Swedish Society for the Deaf and the Swedish Society of Promotion of Hearing Aids, 1.2 million persons would choose to receive subtitled TV programs if they were currently available.

Denmark's national TV network has been on the air since late 1977 with a CEEFAX-based teletext system, broadcasting 55 pages of general information and subtitling.

OTHER DEVELOPMENTS

The rate at which new countries are becoming involved with both teletext and viewdata is too swift to be easily tracked on a country-by-country basis. As of early 1979, tests were being conducted in Belgium, Switzerland, Colombia, Australia, Hong Kong, Singapore and others.

The major problem today is to choose among the competing systems being offered, and to work toward some standardization that would simplify the eventual system to be adopted. A single transmission standard seems unobtainable because of the basic differences in power frequencies, TV channel bandwidths, line scan rates and color encoding systems. However, at least a single display standard could be adopted for the 50 Hz areas around the world, and a similar single display standard for the 60 Hz areas.

It seems inevitable, however, that a number of noncompatible systems will develop and exist side by side.

International Standardization

International standardization of telecommunications equipment is the province of the CCITT (International Telegraph and Telephone Consultative Committee), located in Geneva, Switzerland. The CCITT has established an International Videotex Working Party to investigate standards for videotext display. Because of the differing color television transmission systems around the world (see Chapter 2), the likelihood is that several different standards will have to be adopted.

7

Conclusion

by Efrem Sigel

The developers of a new technology are rarely able to predict how it will ultimately be used. Thomas Edison saw the phonograph as a medium of education and cultural enlightenment, scarcely envisioning the enormous entertainment industry that would grow up around recorded music. Marconi conceived of radio as a medium for point-to-point communication, not for mass broadcasting. The early computer pioneers saw their giant machines as an aid in scientific calculations, not as a ubiquitous feature of modern business and social life.

So it may be with videotext. Except that in this case, the developers may be thinking too grandiosely, overestimating the future application of what is, admittedly, a fascinating development of modern telecommunications. It is well to understand that none of the videotext systems discussed in this book will necessarily succeed on a large scale. The technologies face competition from alternate ways of transmitting information. For videotext to be successful, both the suppliers and users of information would have to change ingrained habits and bear new costs.

OBSTACLES TO TELETEXT

Broadcast videotext is the lowest cost technology and thus superficially the most tempting. After making only the most modest investment, a TV station or network can be on the air with a service that uses the vertical interval of an existing signal to transmit additional information. The expenditure required to create the pages for this service is not trivial, but a network like CBS in the U.S. or the BBC in Britain will have no trouble supporting the cost as part of a wider program in news gathering.

The real question comes down to consumer need for the service. Obviously there are some individuals addicted to news or financial reports, who absolutely must be plugged into a device that feeds them the latest information. But how many are there, and in what way are they willing to support the service by paying for it?

No videotext service can grow unless the cost of the terminal—the adaptor that hooks up to the TV set—is low enough to stimulate demand. But before manufacturers will lower the price, they want to be assured that consumer demand does exist, that the market is attractive enough to make the investment. Hence, there are two parties that must make an investment in broadcast videotext: the broadcaster and the TV manufacturer.

If videotext were the only opportunity to extend the use of the TV set, perhaps manufacturers would rush to embrace it. But clearly, videotext is not the only alternative. Video games that use the TV set are one such use. Video cassette recorder/players and video disc players are another use. Home computers hooked to the TV set are a third possibility. A viewdata-like system, linking the home TV to a central computer, is a fourth. And broadcast videotext is a fifth.

In the world of science fiction or popular science magazines, it is possible to conjure up the vision of the home entertainment/information center — to imagine a world in which every living room is equipped with a combined TV receiver/video recorder/home computer/facsimile transceiver and so on.

Reality is more prosaic. Consumers do not want, nor can they afford, every possible gadget. They will pick and choose. The choice of a video recorder may mean the rejection of a videotext decoder —not consciously, but because the individual is going to spend his time watching old movies or tennis lessons on the tube, not scanning news bulletins. Getting hundreds of thousands, or millions, of consumers to adopt a new piece of equipment takes more than manufacturing and technical know-how. It demands a marketing campaign to make people aware of and disposed to purchase the device.

The slow penetration of teletext in Britain between 1976 and the end of 1978 illustrates how dependent a new technology is on marketing. Barely 15,000 teletext sets were sold in this period. A similar result in the U.S. would probably cause manufacturers to abandon the market posthaste.

There are at least four natural obstacles to the growth of broadcast videotext. One is the established habit of watching TV as a pastime, for entertainment only. Another is the fact that reading for any sustained time off the television screen is uncomfortable. A third is that videotext information is often the type that people like to absorb over the breakfast table or on the train, not in the living room — and the TV set is not portable. And a fourth is that newspapers and magazines have established ways of satisfying information needs at fairly low cost, thus making them difficult to compete with.

The study commissioned by the National Cable Television Association in 1979 (see Chapter 5) shows how ingrained is the attitude that television is for entertainment. A news and financial channel ranked last among the choices of additional services that cable television could provide, even though 39% of the respondents still said such a service would be of interest to them.

OBSTACLES TO VIEWDATA

For viewdata systems that depend on a telephone line to link the home or office TV set to a computer, many of the same sorts of difficulties hold. The TV set manufacturers have to be confident that there is a sizeable market before they agree to build sets incorporating the microelectronic circuits that will convert computer signals into a readable pattern on the TV screen. Even more critical than convincing the set manufacturers is convincing the microprocessor suppliers. Companies like Texas Instruments, National Semiconductor and Fairchild want to turn out their chips by the hundreds of thousands, if not the millions. Only with this volume can prices be made low enough to give videotext terminals the chance to appeal to a wide audience.

There is a special problem involved in the interactive videotext systems, those using the TV as a computer terminal. This is the need for the publisher or information supplier to take an active role in identifying customers and in marketing to them. Probably the biggest pitfall awaiting Prestel in the United Kingdom is that publishers who have pages on the system seem to expect customers to seek them out. Moreover, the bewildering profusion of kinds of information available on Prestel works against any single publisher's establishing a clear identification of his pages on the system. But without such identification, users have no incentive to subscribe in the first place.

The nature of information users is that they know what they are interested in, and they buy publications or watch programs that cater to those interests. Certainly there are some general publications, like the daily newspapers that readers buy without knowing what will be in a given issue. (Even with newspapers, customers know that the sports page or the stock page or the crossword puzzle will hold their interest.) Much of publishing, however, is the identification and satisfying of specialized information needs: the magazine on cars, the book on investing in real estate, the directory of film labs or the newsletter on the oil industry are all publications that have aimed squarely at an identifiable audience.

But an organization that stores information in a viewdata computer has no idea who the individual is who will buy a viewdata terminal for his living room. This is the reason that viewdata's licensees in the U.S. are determined to first offer the service to business and professional markets, before venturing into the consumer market.

A publisher of scientific or technical or business information can use videotext to reach existing customers for his materials, offering the instantaneous feature of computer retrieval for just those organizations that are willing to pay the freight. Such a publisher will want to store information in a system that is custom-tailored to his audience, not filled with hundreds of thousands of pages from a hodgepodge of suppliers.

The most serious error that viewdata proponents can make—and one that appears to have been made in Britain—is to apply an analogy drawn from the world of telephone lines and common carriers to the world of intellectual property. The British Post Office has attempted to create a vast information utility in which everything is available from a single source. This ignores the fact that the publisher of *Time* spends his days trying to distinguish his publication from *Newsweek;* the publisher of *The Times* of London insists on being different from *The Sun*; Moody's wants an identity separate from Standard & Poor's, and so on.

PROMISE OF VIDEOTEXT

These are the considerations that should give pause to enthusiastic converts who believe that videotext will revolutionize information transfer. On the other hand, there is much in videotext technologies that holds undeniable promise. One fundamental advantage is that they make use of communications channels already in place. In the case of broadcast videotext, insertion of teletext pages can be easily accomplished at the time the basic signal is transmitted. The added equipment cost is minor, and broadcasters already reach millions of homes. Nevertheless, broadcast videotext will always remain an add-on service to a broadcasting enterprise whose main activity is entertainment. The real potential for revolution in information retrieval lies with viewdata-type services.

In the case of these systems, making use of the telephone network provides ubiquitous access to homes and offices. Technology has brought down the cost of moving digital information from one place to another, through domestic satellites, packet-switching and other techniques. The cost of computer storage is also falling steadily; more and more information is in machine-readable form.

Thus, electronic dissemination of information is on the verge of important breakthroughs, and two of the missing links are low-cost display terminals (such as an adapted color TV set) and computer software that facilitates easy access to remote data banks. To the extent that viewdata systems provide these missing links, they can be in the forefront of the shift to electronic information retrieval.

Technology makes possible shifts in social behavior, but it does not guarantee such shifts will occur. Acceptance of videotext requires a decision by publishers to supply—and by consumers to accept—information in a new way. It requires economic incentives powerful enough for information suppliers to invest in computer storage and manipulation, and for equipment manufacturers to make and market a new kind of information terminal. It requires cooperation by government agencies or authorities to permit unhindered development of either broadcast or telephone line services, without imposing a straitjacket of regulation, or letting entrenched interests control the new medium. And it requires, finally, enough confidence in the future of this technology for its developers to nurture it for many years: the changes it entails in established ways of doing things are so profound that they will never take place in a period of months, or even in a year or two.

Should all these things occur, the future of videotext in the 1980s is promising indeed. Should any fail to occur, videotext services may be nothing more than a curious detour on a long evolutionary journey to new forms of communication.

Appendix A

Organizations Involved with Videotext

ALBERTA GOVERNMENT TELEPHONE
10020 100 St.
Edmonton, Alberta, Canada

AMERICAN TELEVISION & COMMUNICATIONS
20 Inverness Place
Englewood, CO 80110

ASAHI BROADCASTING
2-2-48 Ohyodo-Minami
Ohyodo Ku, Osaka 531, Japan

ASSOCIATED PRESS
50 Rockefeller Plaza
New York, NY 10020

AT&T
195 Broadway
New York, NY 10003

BARCO ELECTRONIC NV
TH Sevenslaan 1.6
8500 Kortrijk, Belgium

BARIC COMPUTING SERVICES LTD.
Forest Rd.
Feltham, Middlesex, England

BBC TELEVISION CENTRE (CEEFAX)
Wood Lane,
London W12 7RJ, England

BELL CANADA
Vista Demonstration Centre
25 Eddy St.
Hull, Quebec J8X 2V7, Canada

BONNEVILLE INTER-NATIONAL CORP.
36 S. State St.
Salt Lake City, UT 84111

BREMA (British Radio Equipment Manufacturers' Association)
Twentieth Century House
31 Soho Sq., London W1V 5DG
England

BRITISH COLUMBIA TELEPHONE CO.
3777 Kingsway
Burnaby, Vancouver, Canada

BRITISH POST OFFICE
23 Howland St.
London WI, England

Prestel Headquarters:
Telephone House
Temple Ave.
London EC4 Y OHL, England

BUTLER COX
Morley House
26-30 Holborn Viaduct
London EC1A 2BP, England

CAP-CPP
14 Great James St.
London WC1, England

CBS
51 W. 52 St.
New York, NY 10019

CCETT (Centre Commun d'Etudes de Télévision et Télé-communications)
PTT-TDF
35 Rennes, France

CCITT (International Telegraph & Telephone Consultative Committee)
International Telecommunication Union
Place des Nations
1211 Geneva 20
Switzerland

CHERRY LEISURE
387 High Road
London NW10, England

COMMUNICATIONS RE-SEARCH CENTRE
Canadian Department
of Communications
300 Slater St.
Ottawa K1A OC8, Canada

D.E.R.
Apex House
Twickenham Road
Feltham, Middlesex, England

DECCA RADIO & TV LTD.
15 Ingate Place
London SW8, England

DIGITAL BROADCASTING CO.
1616 Andersen Rd.
McLean, VA 22102

DOW JONES
22 Cortlandt St.
New York, NY 10007

ELECTRONIC INDUSTRIES ASSOC.
2001 Eye St., NW
Washington, DC 20006

EXCHANGE TELEGRAPH
Extel House, East Harding St.
London EC4, England

FEDERAL COMMUNI-CATIONS COMMISSION
1919 M St., NW
Washington, DC 20554

FINANCIAL TIMES
Arthur Street
London EC4R 9AX, England

FINTEL
1 Pudding Lane
London EL4R 8AA, England

F.P. PUBLICATIONS LTD.
444 Royal Trust Tower
Toronto, Canada

GEC RADIO & TELEVISION LTD.
Langley Park
Slough SL3 6DP, England

GENERAL ELECTRIC CO. (GEC)
1 Stanhope Gate
London W1, England

GENERAL TELEPHONE & ELECTRONICS
1 Stamford Forum
Stamford, CT 06904

GERMAN BUNDESPOST
19 Dernburgstrasse 50
West Berlin, W. Germany

GRANADA TV RENTAL LTD.
Ampthill Road
Bedford MK42 9QQ, England

IBA (Independent Broadcasting Authority)
70 Brompton Rd.
London SW3, England

INSAC VIEWDATA
227 Park Ave.
New York, NY 10017

INSTITUT FUR RUNDFUNK-TECHNIK
45 Floriansmühlstrasse 60
Munich, W. Germany

ITT CONSUMER PRODUCTS (UK) LTD.
Maidstone Road, Foots Cray
Sidcup, Kent, England

**KIRBY LESTER ELEC-
TRONICS**
Osborne Mill, Waddington St.
Oldham, Lancashire OL9 6QQ
England

KMOX-TV
1 Memorial Drive
St. Louis, MO 63102

KNIGHT-RIDDER
1 Herald Plaza
Miami, FL 33101

KSL-TV
145 Social Hall Ave.
Salt Lake City, UT 84111

LABGEAR LTD.
Abbey Walk
Cambridge CB1 2RQ, England

LINK ASSOCIATES
215 Park Ave. S.
New York, NY 10003

LOGICA LTD.
31 Foley St.
London W1, England

LONDON STOCK EXCHANGE
London EC2N 1HP, England

**MACDONALD EDUCA-
TIONAL**
Holywell House, Worship St.
London EC2A 2EW, England

MANHATTAN CABLE TV
120 E. 23 St.
New York, NY 10010

**MANITOBA TELEPHONE
SYSTEM**
517 18 St.
Brandon, Manitoba, Canada

MC GRAW-HILL
1221 Ave. of the Americas
New York, NY 10020

**MICROBAND CORP. OF
AMERICA**
655 Third Ave.
New York, NY 10017

MICRO TV
(Sub. of Radio Broadcasting Co.)
3600 Conshohocken Ave.
Philadelphia, PA 19131

MILLS AND ALLEN
15/17 Broadwick St.
London W1V 2AH, England

**MINISTRY OF POSTS AND
TELECOMMUNICATIONS**
20 Avenue de Segur
75700 Paris, France

MITEL CORPORATION
Hwy. 17W Kanata
Ottawa, Ontario, Canada

**NATIONAL ASSOCIATION OF
BROADCASTERS**
1771 N St., NW
Washington, DC 20036

**NATIONAL CABLE
TELEVISION ASSOCIATION**
918 16 St., NW
Washington, DC 20006

NEW YORK TIMES CO.
229 W. 43 St.
New York, NY 10036

NHK (Nippon Hoso Kyokai)
2-2-1 Jinnan
Shibuya-ku, Tokyo, Japan

**NIPPON TELEGRAPH AND
TELEPHONE**
1-1-6, Uchisaiwai-Cho
Chiyoda-Ka
Tokyo 100, Japan

NORPAK, LTD.
Pakenham, Ontario, KOA 2XO
Canada

**NORTHERN TELECOM
CANADA LIMITED**
1600 Dorchester W.
Montreal, Canada

NOS (Nederlandse Omroep
Stichting)
P.O. Box 10
1200 JB Hilversum
The Netherlands

OAK INDUSTRIES
Communications Division
Crystal Lake, IL 60014

**ONTARIO EDUCATIONAL
AND COMMUNICATIONS
AUTHORITY**
P.O. Box 200, Station Q
Toronto, Ontario M4T 2T1
Canada

PHILIPS ELECTRICAL LTD.
Arundel Court, 8 Arundel St.
London WC2, England

**PUBLIC BROADCASTING
SERVICE (PBS)**
475 L'Enfant Plaza S.W.
Washington, DC 20024

PYE LTD.
137 Ditton Walk
Cambridge CB5 8QD, England

RADIO RENTALS
Relay House, Percy St.
Swindon SN2 2BB, England

RADIO BROADCASTING CO.
3600 Conshohocken Ave.
Philadelphia, PA 19131

**RANK RADIO INTER-
NATIONAL**
Northolt Ave., Ernsettle
Plymouth, Devon PL5 2TS
England

RCA
30 Rockefeller Plaza
New York, NY 10020

**REDIFFUSION CONSUMER
ELECTRONICS LTD.**
c/o Rediffusion Ltd.
Stratton House, Stratton St.
London W1, England

REUTERS (North America)
1700 Broadway
New York, NY 10019

**SATELLITE SYNDICATED
SYSTEMS**
P.O. Box 45684
Tulsa, OK 75145

S.E.D. SYSTEMS LTD.
2415 Koyl
Saskatoon, Saskatchewan, Canada

SOFRATEV
124 bis Avenue de Villiers
75017 Paris, France

SONY (UK) LTD.
Pyrene House
Sunbury Cross
Sunbury on Thames, England

SOUTHAM, INC.
321 Bloor E.
Toronto, Ontario, Canada

**STANDARD TELEPHONES
AND CABLES LTD.**
Oakleigh Rd., South
New Southgate, London N11 1HB
England

SVERIGES RADIO
Oxenstiernsgatan 20
S-105
101 Stockholm, Sweden

TELECABLE VIDEOTRON
3700 Losch Blvd.
St. Hubert, Quebec J3Y5T6
Canada

TELECOMPUTING CORP. OF AMERICA
1616 Anderson Rd.
McLean, VA 22102

TELEFUSION LTD.
Telefusion House
Preston New Rd.
Blackpool, England FY4 4QY

TELEPROMPTER CORP.
888 7th Ave.
New York, NY 10019

TEXAS INSTRUMENTS
P.O. Box 2474
Dallas TX 75222

TEXAS INSTRUMENTS LTD.
Manton Lane
Bedford MK41 7PA, England

TDF (Télévision de France)
2127 Rue Barbes
Montrouge 92100, France

THORN CONSUMER ELECTRONICS LTD.
Thorn House
Upper St. Martin's Lane
London WC2, England

TIME INC.
Time & Life Building
New York, NY 10020

TOCOM INC.
Box 47066
Dallas, TX 75247

TORSTAR CORPORATION
1 Wonge St.
Toronto, Ontario M5E1P9
Canada

UNITED PRESS INTERNATIONAL
220 E. 42 St.
New York, NY 10017

VIACOM
1211 Ave. of the Americas
New York, NY 10036

VIEWDATA CORP. OF AMERICA
c/o Knight-Ridder
1 Herald Plaza
Miami, FL 33101

WARNER CABLE
10 Rockefeller Plaza
New York, NY 10020

WTCG-TV
1018 W. Peachtree St.
Atlanta, GA 30309

ZENITH RADIO CORP.
1000 Milwaukee Ave.
Glenview, IL 60025

Appendix B

Prestel Information Providers

ABC TRAVEL GUIDES
World Timetable Centre
Dunstable, Beds., LU6 3EB

ACCESS CREDIT CARDS
Chartwell House, Chartwell Sq.
Southend SS2 5S

AGRA EUROPE
16 Lonsdale Gardens
Tunbridge Wells, Kent

ADVISORY UNIT FOR COM-PUTER BASED EDUCATION
19 St. Albans Rd. East
Hatfield, Herts.

APPLIED VIEWDATA SYSTEMS
Yeoman House
76 St. James's La.
London N10 3RD

ARAMBY AGENCIES
175 Piccadilly
London W1V 9DB

ASLIB
36 Bedford Row
London WC1R 4JH

AUSTIN KNIGHT
20 Soho Sq.
London W1A 1DS

THE BARCLAYS GROUP BARCLAYCARD
Juxon House
94 St. Paul's Churchyard
London EC4M 8EH

BARIC COMPUTING SERVICES
Forest Rd.
Feltham, Middlesex, TW13 7EJ

BENN BROTHERS
102-104 High St.
Croyden, Surrey CRO 1ND

BIBLE SOCIETY
146 Queen Victoria St.
London EC4V 4BX

BIRMINGHAM PUBLIC LIBRARIES
Quick Reference Department
Central Library
Birmingham B3

BRITANNIA GROUP OF INVESTMENT COMPANIES
3 London Wall Buildings
London EC2M 5QL

BRITISH AIRWAYS
West London Terminal
Cromwell Rd.
London SW7 4ED

BRITISH INSURANCE ASSOCIATION
Aldermary House, Queen St.
London EC4P 4JD

BRITISH LIBRARY
Blaise, 7 Rathbone St.
London W1P 2AL

BRITISH MEDICAL ASSOCIATION (BMA)
Central Advisory Committee
Tavistock Square
London WC1H 9JP

BRITISH OXYGEN (BOC LTD.)
Hammersmith House
Hammersmith, London W6 9DX

BRITISH RAIL
222 Marylebone Rd.
London NW1 6JJ

**BRITISH TOURIST
AUTHORITY**
239 Old Marylebone Rd.
London NW1

BUCKMASTER & MOORE
The Stock Exchange
London EC2P 2JT

BUPA (The British United Provident Association Ltd.)
Provident House, Essex St.
London WC2R 3AX

**BUSINESS TRANSFER
VIEWDATA**
BTV PO Box 32
Cheltenham GL54 4HH

BUTLER COX
Morley House
26 Holborn Viaduct
London EC1 2BP

**CANCER RESEARCH
CAMPAIGN**
2 Carlton House Terrace
London SW1 5AR

CAREERDATA
Yeoman House
St. James La.
London N10 3RD

**THE CAXTON PUBLISHING
CO.**
72-90 Worship St.
London EC2A 2EN

**CENTRAL OFFICE OF
INFORMATION**
Hercules Rd.
London SE1 7DU

**CENTRAL STATISTICAL
OFFICE**
Great George St.
London SW1P 3AQ

**COMMUNICATIONS
STUDIES & PLANNING**
21 Great Titchfield St.
London W1

CONSUMERS' ASSOCIATION
14 Buckingham St.
London WC2N 6DS

THOMAS COOK
Thorpe Wood
Peterborough PE3 6SB

COTTON OUTLOOK
GO1-GO7 Cotton Exchange
Building
Liverpool L39JR

**COUNCIL FOR EDUCATIONAL
TECHNOLOGY**
3 Devonshire St.
London W1N 2BA

CRONER PUBLICATIONS
Tolworth Tower, Ewell Rd.
Surbiton, Surrey KT6 7EY

**DATASTREAM INTER-
NATIONAL**
9/12 King St.
London EC2

**DEPARTMENT OF THE
ENVIRONMENT**
2 Marsham St.
London SW1P 3EB

DEPARTMENT OF INDUSTRY
Room 725, Sanctuary Buildings
16-20 Great Smith St.
London SW1

**DEPARTMENT OF PRICES &
CONSUMER PROTECTION**
Room 725, Sanctuary Buildings
16-20 Great Smith St.
London SW1

DEPARTMENT OF TRADE
Room 725, Sanctuary Buildings
16-20 Great Smith St.
London SW1

EASTEL
Eastern Counties Newspapers
Limited
Prospect House, Rouen Rd.
Norwich NR1 1RE

THE ECONOMIST
25 St. James's St.
London SW1 1HG

ENGLISH TOURIST BOARD
4 Grosvenor Gardens
London, SW1W ODU

**EXCHANGE & MART
PUBLISHING**
Robert Rogers House
Poole, Dorset BH15 1LU

EXTEL SPORT
Extel House, East Harding St.
London EC4P 4HB

FAMILY LIVING
National Magazine House
72 Broadwick St.
London W1

FINTEL
1 Pudding La.
London EC3R 8AA

**THE GENERAL ELECTRIC
COMPANY**
GEC Viewdata Systems
Kemble House, Kemble St.
London WC2B 4AJ

**GILTSPUR BULLENS
FREIGHT**
Elstree Way
Boreham Wood, Herts.

**GKN DISTRIBUTORS
FASTENER DIVISION**
Lichfield Rd.
Tamworth, Staffs. B79 7TF

THE GOOD BOOK GUIDE
Braithwaite & Taylor Ltd.
PO Box 28
London SW11 4BT

GOOD FOOD GUIDE
14 Buckingham St.
London, WC2N 6DS

**GRAND METROPOLITAN
HOTELS**
7 Stratford Pl.
London W1

**GRAND METROPOLITAN
SYSTEMS**
Oxford House, Oxford Rd.
Uxbridge UB8 1UN

**GREATER LONDON
COUNCIL**
CCS (Prestel), County Hall
London SE1 7PB

W. GREENWELL
Bow Bells House
Bread St.
London EC4M 9EL

GUINNESS SUPERLATIVES
2 Cecil Court, London Rd.
Enfield, Middlesex

**HARTE-HANKS COMMUNI-
CATIONS**
Box 269
San Antonio, TX 78291
USA

HAYMARKET PUBLISHING
54-62 Regent St.
London W1A 4YU

**HEALTH AND SAFETY
EXECUTIVE**
Baynards House
1 Chepstow Place
London W2 4TF

**HENDERSON
CROSTHWAITE & CO.**
194-200 Bishopsgate
London EC2M 4LL

HOARE GOVETT
Atlas House, 1 King St.
London EC2

HOBSONS PRESS
Bateman St.
Cambridge CB2 1LZ

HOLIDAY WHICH?
Consumers' Association
14 Buckingham St.
London WC2N 6DS

HORIZON
214 Broad St.
Birmingham

IBM UK
28 The Quadrant
Richmond, Surrey TW9 1DW

ICI PLASTICS DIVISION
Bessemer Rd.
Welwyn Garden City
Herts. AL7 1HD

**INDUSTRIAL EXCHANGE &
MART**
Robert Rogers House
New Orchard
Poole, Dorset

INFOLEX SERVICES
20 Grange Rd. Wickham Bishops
Witham, Essex

INSPEC
Station House, Nightingale Rd.
Hitchin, Herts.

INVESTORS CHRONICLE
Greystoke Place, Fetter La.
London EC4A 1ND

IPC VIEWDATA
79-80 Blackfriars' Rd.
London SE1 8HN

ISI (SCITEL)
132 High St.
Uxbridge, Middlesex

**KNIGHT COMPUTER
SERVICES**
14 Old Park La.
London W1Y 4NL

**LANGTON INFORMATION
SYSTEMS**
74 Newman St.
London W1P 3LA

LINDUSTRIES
Trevor House, 100 Brompton Rd.
London SW3 1EL

**LINK HOUSE COMMUNI-
CATIONS**
Robert Rogers House
New Orchard
Poole BH15 1LU

**LOANS BUREAU, CIPFA
SERVICES**
232 Vauxhall Bridge Rd.
London SW1

**MACDONALD
EDUCATIONAL BOOKS**
72-90 Worship St.
London EC2A 2EW

MACLAREN PUBLISHERS
PO Box 109
Croydon CR9 1QH

MARS
266 Bath Rd.
Slough, Berks. SL1 4EB

McCORQUODALE BOOKS
3 Bedford Row
London WC1R 4BU

METEOROLOGICAL OFFICE
London Rd.
Bracknell, Berks.

METROTECH
Wyvern Way, Rockingham Rd.
Uxbridge

**MILLS & ALLEN COMMUNI-
CATIONS**
Broadwick House, Broadwick St.
London W1V 2AH

MONEY MANAGEMENT
Greystoke Place, Fetter La.
London EC4A 1ND

MONEY WHICH?
Consumers' Association
14 Buckingham St.
London WC2N 6DS

MORGAN-GRAMPIAN
30 Calderwood St.
London SE18 6QH

MOTORING WHICH?
Consumers' Association
14 Buckingham St.
London WC2N 6DS

THE MULTIPLE SCLEROSIS SOCIETY
4 Tachbrook St.
London SW1V 1SJ

NAPIER COLLEGE
Colinton Rd.
Edinburgh EH10 5DT

NATIONAL ASSOCIATION OF CITIZENS' ADVICE BUREAUX
110 Drury La.
London WC2B 5SW

NATIONAL BUILDING AGENCY
7 Arundel St.
London WC2R 3DZ

NATIONAL CONSUMER COUNCIL
18 Queen Anne's Gate
London SW1

NATIONAL SAVINGS
Charles House
375 Kensington High St.
London W14 8SD

NAVAC (National Audio-Visual Aids Centre)
254 Belsize Rd., London NW6

NETWORK DATA (PROPERTY SERVICES)
Imperial House, 15-19 Kingsway
London WC2B 6UX

NEWBOOK NEWS
Viewdata Services
Link House Communications Ltd.
Robert Rogers House
New Orchard, Poole, BH15 1LU

NEW YORK TIMES
London Bureau Ltd.
76 Shoe La.
London EC4A 3JB

NORWICH UNION INSURANCE
4 Surrey St.
Norwich NOR 88A

OFFICE OF FAIR TRADING
15-25 Bream's Buildings
London EC4A 1PR

OPPORTUNITIES
Pembroke House, Wellesley Rd.
Croydon, Surrey

OPTICAL INFORMATION COUNCIL
Walter House, 418-422 Strand
London WC2R OPB

PA INTERNATIONAL MANAGEMENT CONSULTANTS
Rutland House, Rutland Gardens
London SW7 1BY

PAN AMERICAN WORLD AIRWAYS
Heathrow Airport
Hounslow, Middlesex

PEAT, MARWICK, MITCHELL & CO.
1 Puddle Dock, Blackfriars
London EC4V 3PD

PHILLIPS & DREW
Lee House
London Wall EC2Y 5AP

PIRA
Randalls Rd.
Leatherhead, Surrey

**POST OFFICE TELE-
COMMUNICATIONS**
Room 415, Cheapside House
138 Cheapside
London EC2V 6JH

**PROFESSIONAL AND
EXECUTIVE RECRUITMENT**
4-5 Grosvenor Place
London SW1X 7SB

QANTAS
500 Chiswick High Rd.
London W4 5RW

QUOTEL
83 Clerkenwell Rd.
London EC1

REUTERS
85 Fleet St.
London EC4P 4AJ

ROYAL AUTOMOBILE CLUB
83-85 Pall Mall
London SW1 5HW

**ROYAL INSTITUTION OF
CHARTERED SURVEYORS**
12 Great George St.
London SW1P 3AD

**ROYAL NATIONAL INSTITUTE
FOR THE DEAF**
105 Gower St.
London WC1E 6AH

ST. ALBANS COLLEGE
29 Hatfield Rd.
St. Albans, Herts.

SAVE THE CHILDREN FUND
157 Clapham Rd.
London SW9 OPT

E.B. SAVORY MILLN & CO.
20 Moorgate
London EC2R 6AQ

**SCOTTISH COUNCIL FOR
EDUCATIONAL
TECHNOLOGY**
16 Woodside Terrace
Glasgow G3 7XN

SEALINK
163-203 Eversholt St.
London NW1 1BG

SEASPEED
British Rail Hovercraft Ltd.
50 Liverpool St.
London EC2M 7QH

**SHIPSTATS WEEKLY
REPORT**
35 Andrew's House
Barbican, London W8

SPORTS COUNCIL
70 Brompton Rd.
London SW3 1EX

**STANDARD TELEPHONES
AND CABLES**
190 The Strand
London WC2R 1DU

STOCK EXCHANGE
London EC2N 1HP

**SUTTON, LONDON
BOROUGH OF**
Sutton Libraries and Arts Services
Central Library, St. Nicholas Way
Sutton, Surrey

TELEMACHUS
PO Box 86
Aylesbury, Bucks

TJAEREBORG
7-8 Conduit St.
London W1

**TRAMWAY MUSEUM
SOCIETY**
412 Kings Rd. Higher Hurst
Ashton-under-Lyne, Lancs.

**TRANSPORT AND ROAD
RESEARCH LABORATORY,
DoE/DTp**
Crowthorne, Berks. RG11 6AU

**UNITED KINGDOM
CHEMICAL INFORMATION
SERVICE (UKCIS)**
The University
Nottingham NG7 2RD

**UNIVERSITIES CENTRAL
COUNCIL ON ADMISSIONS
(UCCA)**
PO Box 28
Cheltenham, Glos. GL50 1HY

**VIEWDATA MARKETING
SERVICES**
351 Oxford St.
London W1R 1FH

VIEWTEL 202
The Birmingham Post & Mail Ltd.
28 Colmore Circus
Queensway, Birmingham B4 6AX

**WALTHAM FOREST,
LONDON BOROUGH OF**
Central Library, High St.
Walthamstow, London E17 7JN

WATFORD COLLEGE
Hempstead Rd.
Watford, Herts.

WEMSEC
Circle House South
65/67 Wembley Hill Rd.
Wembley, Middlesex

WHICH?
14 Buckingham St.
London WC2N 6DS

**YORKSHIRE POST
NEWSPAPERS**
Wellington St.
Leeds LS1 1RF

Index

About the Authors

Efrem Sigel is editor in chief of Knowledge Industry Publications, Inc. He is the author of *The Kermanshah Transfer,* a novel, and of *Crisis: The Taxpayer Revolt and Your Kids' Schools.* He is a graduate of Harvard College and Harvard Business School, and has been a teacher and a Peace Corps volunteer.

Colin McIntyre, editor of CEEFAX, was the first journalist to participate in the development of the BBC's teletext service. He has been associated with the BBC since 1952, where his positions have included United Nations correspondent, chief publicity officer for BBC Television and Programme Promotions Executive for BBC Television. He is a graduate of Harvard University and the Open University.

Max Wilkinson is the electronics correspondent of *The Financial Times* of London. Previous positions were as education correspondent of the London *Daily Mail* and editor of the weekly newspaper *The Teacher.* He holds a degree in mechanical sciences and English from Cambridge University.

Joseph Roizen, president of Telegen, is a television engineer with a special interest in the field of international television. He has made a worldwide study of teletext systems, visiting research centers in the U.K., France, Germany, Japan and other countries. He is currently the SMPTE liaison member on the EIA/BTS Teletext Subcommittee and is engaged in evaluating various proposals for a U.S. standard.

Other Titles from Knowledge Industry Publications

Who Owns the Media: Concentration of Ownership in the Mass Communications Industry
edited by Benjamin M. Compaine
370 pages   hardbound $24.95

U.S. Book Publishing Yearbook and Directory, 1979-80
edited by Terry S. Mollo
186 pages   softcover $35.00

The Newspaper Industry in the 1980s: An Assessment of Economics and Technology
by Benjamin M. Compaine
250 pages (approx.)   hardcover $24.95

The Book Industry in Transition: An Economic Analysis of Book Distribution and Marketing
by Benjamin M. Compaine
235 pages   hardcover $24.95

Practical Video: The Manager's Guide to Applications
by John A. Bunyan, James C. Crimmins and N. Kyri Watson
203 pages   softcover $17.95

The Video Register, 1979-80 edition
200 pages (approx.)   softcover $34.95

Television and Management: The Manager's Guide to Video
by John Bunyan and James C. Crimmins
154 pages   hardcover $17.95

Trends in Management Development and Education: An Economic Study
by Gilbert J. Black
198 pages   hardcover $24.95

Available from Knowledge Industry Publications, Inc., 2 Corporate Park Drive, White Plains, NY 10604